Lecture Notes in Computer Science 5965

Commenced Publication in 1973
Founding and Former Series Editors:
Gerhard Goos, Juris Hartmanis, and Jan van Leeuwen

Editorial Board

Fabio Massacci   Dan Wallach
Nicola Zannone (Eds.)

Engineering Secure Software and Systems

Second International Symposium, ESSoS 2010
Pisa, Italy, February 3-4, 2010
Proceedings

 Springer

Volume Editors

Fabio Massacci
Università di Trento, Dipartimento Ingegneria e Scienza dell'Informazione
Via Sommarive 14, 38050 Povo (Trento), Italy
E-mail: Fabio.Massacci@unitn.it

Dan Wallach
Rice University, Department of Computer Science
3122 Duncan Hall, 6100 Main Street, Houston, TX 77005, USA
E-mail: dwallach@cs.rice.edu

Nicola Zannone
University of Technology, Faculty of Mathematics and Computer Science
Den Dolech 2, 5612 AZ Eindhoven, The Netherlands
E-mail: n.zannone@tue.nl

Library of Congress Control Number: 2009943930

CR Subject Classification (1998): C.2, E.3, D.4.6, K.6.5, J.2

LNCS Sublibrary: SL 4 – Security and Cryptology

ISSN 0302-9743

ISBN 978-3-642-11746-6 Springer Berlin Heidelberg New York

springer.com

© Springer-Verlag Berlin Heidelberg 2010

Typesetting: Camera-ready by author, data conversion by Scientific Publishing Services, Chennai, India
Printed on acid-free paper   SPIN: 12989713   06/3180   5 4 3 2 1 0

Preface

It is our pleasure to welcome you to the proceedings of the Second International Symposium on Engineering Secure Software and Systems.

This unique event aimed at bringing together researchers from software engineering and security engineering, which might help to unite and further develop the two communities in this and future editions. The parallel technical sponsorships from the ACM SIGSAC (the ACM interest group in security) and ACM SIGSOF (the ACM interest group in software engineering) is a clear sign of the importance of this inter-disciplinary research area and its potential.

The difficulty of building secure software systems is no longer focused on mastering security technology such as cryptography or access control models. Other important factors include the complexity of modern networked software systems, the unpredictability of practical development life cycles, the intertwining of and trade-off between functionality, security and other qualities, the difficulty of dealing with human factors, and so forth. Over the last years, an entire research domain has been building up around these problems.

The conference program included two major keynotes from Any Gordon (Microsoft Research Cambridge) on the practical verification of security protocols implementation and Angela Sasse (University College London) on security usability and an interesting blend of research, industry and idea papers.

In response to the call for papers 58 papers were submitted. The Program Committee selected nine papers as research papers (16%), presenting new research results in the realm of engineering secure software and systems. It further selected one industry report, detailing a concrete case study in industry, and eight ideas papers, that the Program Committee judged interesting but not yet mature for a full paper presentation.

Many individuals and organizations contributed to the success of this event. First of all, we would like to express our appreciation to the authors of the submitted papers, and to the Program Committee members and external referees, who provided timely and relevant reviews. Many thanks go to the Steering Committee for supporting this and future editions of the symposium, and to all the members of the Organizing Committee for their tremendous work and for excelling in their respective tasks. The DistriNet research group of the K.U. Leuven did an excellent job with the website and the advertising for the conference. Nicola Zannone did a great job by assembling the proceedings for Springer.

We owe gratitude to ACM SIGSAC/SIGSOFT, IEEE TCSE and LNCS for supporting us in this new scientific endeavor. An honorable mention should be made of EasyChair, which seems to be "the" system for conference management.

Last but not least, we would like to thank all our sponsors, the SecureChange EU Project at the University of Trento for covering some of the financial aspects of the event and the CNR for organizing the symposium.

December 2009

Fabio Massacci
Dan Wallach
Fabio Martinelli

Conference Organization

General Chair

Fabio Martinelli   CNR, Italy

Program Co-chairs

Fabio Massacci   Università di Trento, Italy
Dan Wallach   Rice University, USA

Publication Chair

Nicola Zannone   Eindhoven University of Technology,
                                      The Netherlands

Publicity Chair

Yves Younan   Katholieke Universiteit Leuven, Belgium

Local Arrangements Chair

Adriana Lazzaroni   CNR, Italy

Steering Committee

Jorge Cuellar   Siemens AG, Germany
Wouter Joosen   Katholieke Universiteit Leuven, Belgium
Fabio Massacci   Università di Trento, Italy
Gary McGraw   Cigital, USA
Bashar Nuseibeh   The Open University, UK
Daniel Wallach   Rice University University, USA

Program Committee

Juergen Doser   IMDEA, Spain
Manuel Fähndrich   Microsoft Research, USA
Michael Franz   UC Irvine, USA
Dieter Gollmann   Hamburg University of Technology, Germany
Jan Jürjens   Open University, UK
Seok-Won Lee   University of North Carolina Charlotte, USA

Antonio Maña   University of Malaga, Spain
Robert Martin   MITRE, USA
Mattia Monga   Milan University, Italy
Fabio Massacci   Università di Trento, Italy
Haris Mouratidis   University of East London, UK
Gunther Pernul   Universitat Regensburg, Germany
Samuel Redwine   James Madison University, USA
David Sands   Chalmers University of Technology, Sweden
Riccardo Scandariato   Katholieke Universiteit Leuven, Belgium
Ketil Stølen   Sintef, Norway
Jon Whittle   Lancaster University, UK
Mohammad Zulkernine   Queen's University, Canada
Neeraj Suri   Technische Universität Darmstadt, Germany
Yingjiu Li   Singapore Management University, Singapore
Hao Chen   UC Davis, USA
Richard Clayton   Cambridge University, UK
Eduardo Fernández-Medina   University of Castilla-La Mancha, Spain
Yuecel Karabulut   SAP Office of CTO, USA
Vijay Varadharajan   Macquarie University, Australia
Jungfeng Yang   Columbia University, USA
Daniel Wallach   Rice University University, USA

External Reviewers

Birgisson, Arnar   Ochoa, Martin
Brandeland, Gyrd   Omerovic, Aida
Buyens, Koen   Paleari, Roberto
Ceccato, Mariano   Passerini, Emanuele
Chowdhury, Istehad   Phung, Phu H.
Desmet, Lieven   Pironti, Alfredo
Dinkelaker, Tom   Pujol, Gimena
Drbeck, Stefan   Refsdal, Atle
Fritsch, Christoph   Reisser, Andreas
Fu, Ge   Riesner, Moritz
Gmelch, Oliver   Ruiz, Jose F.
Hedin, Daniel   Scandariato, Riccardo
Heldal, Rogart   Seehusen, Fredrik
Heyman, Thomas   Shahriar, Hossain
Hossain, Shahriar   Solhaug, Bjornar
Joosen, Wouter   Turkmen, Fatih
Koshutanski, Hristo   Wang, Yongge
Larsson, Andreas   Yan, Qiang
Li, Yan   Yang, Xioafeng
Lund, Mass Soldal   Yautsiukhin, Artsiom
Muñoz, Antonio   Yskout, Koen
Neuhaus, Stephan

Table of Contents

Session 4. Policy Verification and Enforcement II

Session 5. Secure System and Software Development I

Session 6. Secure System and Software Development II

BuBBle: A Javascript Engine Level Countermeasure against Heap-Spraying Attacks

Francesco Gadaleta, Yves Younan, and Wouter Joosen

IBBT-Distrinet, Katholieke Universiteit Leuven, 3001, Leuven Belgium*

Abstract. Web browsers that support a safe language such as Javascript are becoming a platform of great interest for security attacks. One such attack is a heap-spraying attack: a new kind of attack that combines the notoriously hard to reliably exploit heap-based buffer overflow with the use of an in-browser scripting language for improved reliability. A typical heap-spraying attack allocates a high number of objects containing the attacker's code on the heap, dramatically increasing the probability that the contents of one of these objects is executed. In this paper we present a lightweight approach that makes heap-spraying attacks in Javascript significantly harder. Our prototype, which is implemented in Firefox, has a negligible performance and memory overhead while effectively protecting against heap-spraying attacks.

Keywords: heap-spraying, buffer overflow, memory corruption attacks, browser security.

1 Introduction

Web browsing has become an very important part of today's computer use. Companies like Google™and Yahoo are evidence of this trend since they offer full-fledged software inside the browser. This has resulted in a very rich environment within the browser that can be used by web programmers. However, this rich environment has also lead to numerous security problems such as cross site scripting and cross site request forgeries (CSRF). The browser is often written in C or C++, which exposes it to various vulnerabilities that can occur in programs written in these languages, such as buffer overflows, dangling pointer references, format string vulnerabilities, etc. The most often exploited type of C vulnerability is the stack-based buffer overflow. In this attack, an attacker exploits a buffer overflow in an array, writing past the bounds of the memory allocated for the array, overwriting subsequent memory locations. If the attackers are able to overwrite the return address or a different type of code pointer (such

* This research is partially funded by the Interuniversity Attraction Poles Programme Belgian State, Belgian Science policy, the Research Fund K.U.Leuven, the Interdisciplinary Institute for Broadband Technology and the European Science Foundation (MINEMA network).

F. Massacci, D. Wallach, and N. Zannone (Eds.): ESSoS 2010, LNCS 5965, pp. 1–17, 2010.

as a function pointer), they can gain control over the program's execution flow, possibly redirecting it to their injected code. While stack-based buffer overflows are still an important vulnerability in C programs, they have become harder to exploit due to the application of many different countermeasures such as StackGuard [30], ProPolice [14], ASLR [5], etc. The goal of these countermeasures is to protect areas of potential interest for attackers from being modified or to prevent attackers from guessing where their injected code is located, thus preventing them from directing control flow to that location after they have overwritten such an area of potential interest. Attackers have subsequently focussed on different types of vulnerabilities. One important type of vulnerability is the heap-based buffer overflow. However, due to the changing nature of the heap, these vulnerabilities are notoriously hard to exploit. Especially in browsers, where the heap can look completely different depending on which and how many sites the user has visited, it is hard for an attacker to figure out where in the heap space his overflow has occurred. This makes it hard for attackers to figure out where their injected code is located. The application of countermeasures like ASLR (Address Space Layout Randomization) on the heap has made it even harder to reliably exploit these vulnerabilities. ASLR is a technique by which positions of key data areas in a process's address space, such as the heap, the stack or libraries, are arranged at random positions in memory. All attacks based on the knowledge of target addresses (e.g. *return-to-libc* attacks in the case of randomized libraries or attacks that execute injected *shellcode* in the case of a randomized heap/stack) may fail if the attacker cannot guess the exact target address. Recently a new attack emerged that combines the rich environment found in the browser to facilitate exploits of C vulnerabilities, sometimes resulting in the successful bypass of countermeasures like ASLR that are supposed to protect against these type of vulnerabilities. Heap-spraying attacks use the Javascript engine in the browser to replicate the code they want executed a large amount of times inside the heap memory, dramatically increasing the possibility that a particular memory location in the heap will contain their code. Several examples of heap-spraying attacks have already affected widely used web browsers like Safari, Internet Explorer and Mozilla Firefox.

This paper presents an approach that protects against heap-spraying attacks based on the observation that the attack relies on the fact that the heap will contain homogenous data inserted by the attacker. By introducing diversity in the heap at random locations and by modifying the way that Javascript stores data on the heap at these locations, we can build an effective protection against these exploits at low cost. We implemented this countermeasure in the Javascript engine of Firefox, Tracemonkey[1] The overhead of our approach is very low, measuring the overhead of the countermeasure on a number of popular websites which use a significant amount of Javascript, showed that our approach has an average overhead of 5%. The rest of the paper is organized as follows: Section 2 discusses the problem of heap-based buffer overflows and heap-spraying in more detail, while Section 3.1 discusses our approach and Section 3.2 our prototype implementation. Section 4 evaluates our prototype implementation, while Section 5 compares our approach to related work. Section 6 concludes.

---

[1] In the remainder of the paper we will refer to Spidermonkey, since Tracemonkey is based on the former engine to which it adds native-code compilation, resulting in a massive speed increase in loops and repeated code.

2   Problem Description

2.1   Heap-Based Buffer Overflows

The main goal of a heap-spraying attack is to inject malicious code somewhere in memory and jump to that code to trigger the attack. Because a memory corruption is required, heap-spraying attacks are considered a special case of heap-based attacks. Exploitable vulnerabilities for such attacks normally deal with dynamically allocated memory. A general way of exploiting a heap-based buffer overflow is to overwrite management information the memory allocator stores with the data. Memory allocators allocate memory in chunks. These chunks are located in a doubly linked list and contain memory management information (chunkinfo) and real data (chunkdata). Many different allocators can be attacked by overwriting the chunkinfo.

Since the heap memory area is less predictable than the stack it would be difficult to predict the memory address to jump to execute the injected code.Some countermeasures have contributed to making these vulnerabilities even harder to exploit [35,13].

2.2   Heap-Spraying Attacks

An effective countermeasure against attacks on heap-based buffer overflow is Address Space Layout Randomization (ASLR) [5]. ASLR is a technique which randomly arranges the positions of key areas in a process's address space. This would prevent the attacker from easily predicting target addresses. However, attackers have developed more effective strategies that can bypass these countermeasures. Heap spraying [27] is a technique that will increase the probability to land on the desired memory address even if the target application is protected by ASLR. Heap spraying is performed by populating the heap with a large number of objects containing the attacker's injected code. The act of spraying simplifies the attack and increases its likelihood of success. This strategy has been widely used by attackers to compromise security of web browsers, making attacks to the heap more reliable than in the past [4,8,22,25,32] while opening the door to bypassing countermeasures like ASLR.

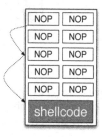

Fig. 1. NOP sled and shellcode appended to the sequence

A heap-spraying attack attempts to increase the probability to jump to the injected code (shellcode). To achieve this, a basic block of NOP[2] instructions is created. The

---

[2] Short for No Operation Performed, is an assembly language instruction that effectively does nothing at all [17].

size of this block is increased by appending the block's contents to itself, building the so called *NOP sled*. Finally `shellcode` is appended to it. A jump to any location within the *NOP sled* will transfer control to the `shellcode` appended at the end. The bigger the sled the higher the probability to land in it and the attack to succeed. A schema of a NOP sled with the shellcode appended to it is provided in Fig.1. The second phase of the attack consists of populating the heap of the browser with a high number of these objects, by using the legal constructs provided by the scripting language supported by the web browser. Figure 2 shows the schema of a heap-spraying attack during heap population. Although in this paper we will refer to heap-spraying the memory space of a web browser, this exploit can be used to spray the heap of any process that allows the user to allocate objects in memory. For instance Adobe Reader has been affected by a heap-spraying vulnerability, by which malicious PDF files can be used to execute arbitrary code [26]. Moreover, heap-spraying is considered an unusual security exploit since the action of spraying the heap is considered legal and permitted by the application. In our specific scenario of a heap-spraying attack in a web browser, memory allocation may be the normal behavior of benign web pages. A web site using AJAX (Asynchronous Javascript And XML) technology, such as facebook.com or plain Javascript such as economist.com, ebay.com, yahoo.com and many others, would seem to spray the heap with a large number of objects during their regular operation. Since a countermeasure should not prevent an application from allocating memory, heap-spraying detection is a hard problem to solve.

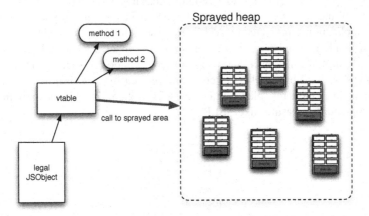

Fig. 2. A heap-spraying attack: heap is populated of a large number of $NOP - shellcode$ objects. The attack may be triggered by a memory corruption. This could potentially allow the attacker to jump to an arbitrary address in memory. The attack relies on the chance that the jump will land inside one of the malicious objects.

Because the structure of the heap depends on how often the application has allocated and freed memory before the spraying attack, it would be difficult to trigger it without knowing how contents have been arranged. This would reduce the attack to a guess of the correct address the malicious object has been injected to. But by using a client-side scripting language, such as Javascript, it is also possible to create the ideal circumstances for such an attack and arrange the heap to get the structure desired by the attacker, as described in [28,11].

Fig. 3 shows an example of a typical heap-spraying attack in Javascript.

```
1. var sled;
2. var spraycnt = new Array();
3. sled = <NOP_instruction>;
4. while(sled.length < _size_)
5.    {
6.        sled+=sled;
7.    }
8. for(i=0; i< _large_; i++)
9.    {
10.       spraycnt[i] = sled+shellcode;
11.    }
```

Fig. 3. A Javascript code snippet to perform a basic heap-spraying attack usually embedded in a HTML web page

3 BuBBle: Protection against Heap-Spraying

In this section we describe our approach to prevent the execution of shellcode appended to a NOP sled, when a heap-spraying attack and a memory corruption have occured. Our general approach is described in Section 3.1 and our implementation is discussed in Section 3.2.

3.1 Approach

An important property of a heap-spraying attack is that it relies on homogeneity of memory. This means that it expects large parts of memory to contain the same information (i.e., it's nop-shellcode). It also relies on the fact that landing anywhere in the nopsled will cause the shellcode to be executed. Our countermeasure breaks that assumption by introducing diversity on the heap, which makes it much harder to perform a heap-spraying attack. The assumption is broken by inserting special interrupting values in strings at random positions when the string is stored in memory and removing them when the string is used by the application. These special interrupting values will cause the program to generate an exception when it is executed as an instruction. Because these special values interrupt the strings inside the memory of the application, the attacker can no longer depend on the nopsled or even the shellcode being intact. If these values were placed at fixed locations, the attacker could attempt to bypass the code by inserting jumps over specific possible locations within the code. Such an attack, however is unlikely, because the attacker does not know exactly where inside the shellcode control has been transferred. However, to make the attack even harder, the special interrupting values are placed at random locations inside the string. Since an attacker does not know at which locations in the string the special interrupting values are stored, he can not jump over them in his nop-shellcode. This lightweight approach thus makes heap-spraying attacks significantly harder at very low cost.

We have implemented this concept in the Javascript engine of Firefox, an opensource web browser. The internal representation of Javascript strings was changed in order to add the interrupting values to the contents when in memory and remove them properly whenever the string variable is used or when its value is read. Interruption is regulated by a parameter which can be chosen at browser build time. We have chosen this to be 25 bytes, which is the smallest useful shellcode we found in the wild [33]. This parameter will set the interval at which to insert the special interrupting values. Given the length n of the string to transform i ($i = \lceil \frac{n}{25} \rceil$) intervals are generated (with $n > 25$). A random value is selected for each interval. These numbers will represent the positions within the string to modify. The parameter sets the size of each interval, thus the number of positions that will be modified per string. By choosing a lower value for the parameter the amount of special interrupting values that are inserted will be increased. Setting the size of each interval to the length of the smallest shellcode does not guarantee that the positions will be at distance of 25 bytes. It may occur that a position p is randomly selected from the beginning of its interval i_p and the next position q from the end of its interval i_q. In this case $(q - p)$ could be greater than 25, allowing the smallest shellcode to be stored in between. However heap-spraying attacks are based on large amounts of homogeneous data, not simply on inserting shellcode. Thus being able to insert this shellcode will not simply allow an attacker to bypass this approach. When the characters at random positions are changed, a support data structure is filled with *metadata* to keep track of original values and where in the string they are stored. The modified string is then stored in memory. Whenever the string variable is used, the engine will perform an inverse function, to restore the string to its original value to the caller. This task is achieved by reading the metadata from the data structure bound to the current Javascript string and replacing the special interrupting values with their original values on a copy of the contents of the string. With this approach different strings can be randomized differently, giving the attacker even less chances to figure out the locations of the special values in the string. Because each string variable stays modified as long as it is stored in memory and a copy of this string variable is only restored to its original value when the application requests access to that a string. When the function processing the string stores the result back to memory, the new string is again processed by our countermeasure. If the function discards the string, it will simply be freed. Moreover the Javascript engine considered here implements strings as immutable type. This means that string operations do not modify the original value. Instead, a new string with the requested modification is returned.

3.2 Implementation

In this section we discuss the implementation details of our countermeasure. It has been implemented on Mozilla Firefox (Ver. 3.7 Beta 3) [15], a widely used web browser and its ECMA-262-3-compliant engine, Tracemonkey (Ver. 1.8.2)[18].

An attacker performing a heap-spraying attack attempts to arrange a contiguous block of values of his choice in memory. This is required to build a sled that would not be interrupted by other data. To achieve this, a monolithical data structure is required.

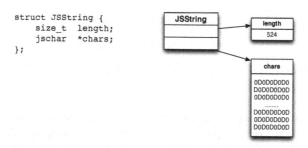

```
struct JSString {
    size_t  length;
    jschar  *chars;
};
```

Fig. 4. Spidermonkey's JSString type is considered a threat for a heap-spraying attack since member *chars* is a pointer to a vector of size $(length + 1)*$ *sizeof(jschar)*

Javascript offers several possibilities to allocate blocks in memory. The types supported by Spidermonkey are numbers, objects and strings. An overview about how the Javascript engine represents Javascript objects in memory is given in [19].

The string type represents a threat and can be used to perform a potentially dangerous heap-spraying. Figure 4 depicts what a JSString, looks like. It is a data structure composed of two members: the length member, an integer representing the length of the string and the chars member which points to a vector having byte size (length + 1) * sizeof(jschar). When a string is created, chars will be filled with the real sequence of characters, representing that contiguous block of memory that the attacker can use as a sled. We have instrumented the JSString data structure with the fields needed for BuBBle to store the metadata: a flag *transformed* will be set to 1 if the character sequence has been transformed and an array *rndpos* is used to store the random positions of the characters that have been modified within the sequence.

Our countermeasure will save the original value of the modified character to *rndpos*, change its value (at this point the string can be stored in memory) and will restore the original value back from *rndpos* whenever the string is read.

This task is performed respectively by two functions: js_Transform(JSString*) and js_Restore(JSString*). The value of the character to modify is changed to the 1-byte value 0xCC. This is the assembly language instruction for x86 processors to

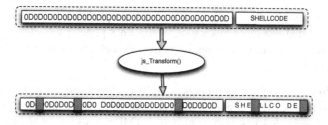

Fig. 5. Representation of the transformed string in memory: characters at random positions are changed to special interrupting values. The potential execution of the object's contents on the heap would be interrupted by the special value.

	0	1	2	3	4	5		21	22
rndpos =	4	"0"	28	"D"	52	"0"	...	507	"E"

Fig. 6. How metadata is stored to array $rndpos$: index i within the array contains the value of the position in the string; index $(i + 1)$ contains its original value

generating a software breakpoint. If a heap-spraying attack was successfully triggered and the byte 0xCC at a random position was executed, an interrupt handler is called to detect and report the attack. The web browser could expose an alert popup to encourage the user close the application and notify the browser vendor of the detected issue. The number of characters to randomize depends on the *degree* parameter. This parameter was chosen based on the length of the smallest shellcode found (to date, 25 bytes long[3]), but can be tuned to select the level of security and the overhead that will be introduced by the countermeasure. If $size$ is the length of the string to transform, the number of intervals is given by $\lceil \frac{size}{24} \rceil$. A random value for each interval will be the position of the character that will be changed. For performance reasons we generate 50 random values in a range between $(0, 24)$ at program startup and use these values as offsets to add to the first index of each interval to compute the random position within that interval. The random values are regenerated whenever the Garbage Collector reclaims memory. This prevents the attacker from learning the values over time as they may already have changed. The value of the ith random position is stored at $rndpos[2i]$, while the original value of the ith character is stored at $rndpos[2i+1]$ (Fig. 6). Function js_Transform(str) will use the values stored in the str→$rndpos[]$ array to restore the string to its original value.

4   Evaluation

In Section 4.1 we discuss our performance overhead, while in Section 4.2 we report an analytical study of the memory overhead in the worst case[4].

All benchmarks were performed on an Intel Core 2 Duo 2Ghz, 4GB RAM, running Debian Gnu/Linux.

4.1   Performance Benchmarks

To measure the performance overhead of BuBBle we performed two types of benchmarks. We collected results of macrobenchmarks on 8 popular websites and accurate timings of microbenchmarks running SunSpider, Peacekeeper Javascript Benchmark, and V8 Benchmark Suite.

Macrobenchmarks: To collect timings of BuBBle's overhead in a real life scenario, we run a performance test similar to the one used to measure the overhead of Nozzle [23]. We downloaded and instrumented the HTML pages of eight popular web sites

---

[3] The smallest setuid and execve shellcode for GNU/Linux Intel x86 to date can be found at
http://www.shell-storm.org/shellcode/files/shellcode-43.php
[4] With *worst case* we mean the case where it is guaranteed that the smallest shellcode cannot be stored on the heap without being interrupted by random bytes.

by adding the Javascript `newDate()` routine at the beginning and the end of the page, and computed the delta between the two values. This represents the time it takes to load the page and execute the Javascript. Since the browser caches the contents of the web page, that value will be close to how long it takes to execute the Javascript. We then ran the benchmark 20 times for each site, 10 times with BuBBle disabled and 10 times with BuBBle enabled. Table 1 shows that the average performance overhead over these websites is 4.8%.

Table 1. Performance overhead of BuBBle in action on 8 popular web sites

Site URL	Load (ms)	Load(ms) BuBBle	Perf. overh.
economist.com	17304	18273	+5.6%
amazon.com	11866	12423	+4.7%
ebay.com	7295	7601	+4.2%
facebook.com	8739	9167	+4.9%
maps.google.com	15098	15581	+3.2%
docs.google.com	426	453	+6.3%
cnn.com	12873	13490	+4.8%
youtube.com	12953	13585	+4.9%
Average			**+4.82**

Microbenchmarks: Microbenchmarks, which allow us to better assess the overheads introduced by BuBBle in different situations were also performed. These microbenchmarks were performed by running three different benchmarks: the SunSpider Javascript Benchmarks [31], the Peacekeeper benchmarks [9] and the V8 benchmarking suite [16].

SunSpider: SunSpider is used by Mozilla Firefox to benchmark the core Javascript language only, without the DOM or other browser dependent APIs. The tests are divided into multiple domains: testing things such as 3D calculations, math, string operations, etc.. Table 2 contains the runtime in milliseconds of running the various benchmarks that are part of SunSpider. The results for each domain are achieved by performing a number of subtests. However for most domains the overhead of the subsets is close to 0%. Thus, to save space in Table 2, we have removed the results of the subtests and simply included the results of the domain (which is sum of all the subtests). However, because we modify the way strings are represented in memory and do a number of transformations on strings, we have included the subtests which test the performance of string operations.

The results in Table 2 show that the overhead for BuBBle in areas other than string manipulation are negligible. The overheads for string operations on the other hand vary from 3% to 27%. This higher overhead of 27% for base64 is due to the way the base64 test is written: the program encodes a string to base64 and stores the result. When the program starts, it generates a character by adding getting a random number, multiplying the number by 25 and adding 97. This character is converted to a string and added to an existing string. This is done until a string of 8192 characters is created. Then to do the encoding, it will loop over every 3rd character in a string and then perform the

Table 2. Microbenchmarks performed by SunSpider Javascript Benchmark Suite

Test		Runtime(ms)	BuBBle Runtime (ms)	Perf. overh.
3d		568.6ms +/- 1.4%	569.0ms +/- 1.2%	**+0.17%**
bitops		66.4ms +/- 1.8%	67ms +/- 1.8%	**+0.89%**
controlflow		13.8ms +/- 1.9%	14.0ms +/- 1.6%	**+1.44%**
math		63.2ms +/- 1.0%	63.6ms +/- 1.7%	**+0.62%**
regexp		84.2ms +/- 2.0%	84.4ms +/- 2.9%	**+0.23%**
string				
	base64	74.8ms +/- 2.9%	102.2ms +/- 1.9%	+27.3%
	fasta	280.0ms +/- 1.5%	283.4ms +/- 0.7%	+1.24%
	tagcloud	293.2ms +/- 2.6%	299.6ms +/- 0.8%	+2.20%
	unpack-code	352.0ms +/- 0.8%	363.8ms +/- 3.1%	+3.24%
	validate-input	119.8ms +/- 2.4%	132.2ms +/- 1.0%	+9.30%
		1119.8ms +/- 0.9%	1181.2ms +/- 1.0%	**+5.19%**

encoding of those three characters to 4 base64 encoded characters. In every iteration of the loop, it will do 7 accesses to a specific character in the original string, 4 access to a string which contains the valid base64 accesses and finally it will do 4 += operations on the result string. Given that our countermeasure will need to transform and restore the string multiple times, this causes a noticeable slowdown in this application.

Table 3. Peacekeeper Javascript Benchmarks results (the higher the better)

Benchmark	Score	BuBBle Score	Perf. overh.
Rendering	1929	1919	+0.5%
Social Networking	1843	1834	+0.5%
Complex graphics	4320	4228	+2.2%
Data	2047	1760	+14.0%
DOM operations	1429	1426	+0.2%
Text parsing	1321	1298	+2.0%
Total score	1682	1635	**+2.8**

Peacekeeper: Peacekeeper is currently used to tune Mozilla Firefox. It will assign a score based on the number of operations performed per second. The results of the Peacekeeper benchmark are located in Table 3: for most tests in this benchmark, the overhead is negligible, except for the Data test which has an overhead of 14%. The Data test is a test which will do all kinds of operations on an array containing numbers and one test which performs operations on an array containing strings of all possible countries in the world. The operations on the strings containing all possible countries are what contribute to the slow down in this benchmark: whenever a country is used, the string is restored, whenever one is modified the resulting new string is transformed.

V8: The V8 Benchmark Suite is used to tune V8, the Javascript engine of Google Chrome. The scores are relative to a reference system (100) and as with Peacekeeper,

Table 4. V8 Benchmark Suite results (the higher the better)

Benchmark	Score	BuBBle Score	Perf. overh.
Richards	151	143	+5.6%
DeltaBlue	173	167	+3.6%
Crypto	110	99.6	+10.4%
Ray Trace	196	193	+1.5%
EarlyBoyer	251	242	+3.7%
RegExp	174	173	+0.6%
Splay	510	501	+1.8%
Total score	198	193	**+2.6**

the higher the score, the better. Again, most overheads are negligible except for Crypto, which has an overhead of 10.4%. Crypto is a test encrypts a string with RSA. To encrypt this string the application does a significant number of string operations, resulting in transformation and restoration occurring quite often.

These benchmarks show that for string intensive javascript applications that do little else besides run string operations, the overhead can be significant, but not a show stopper. In all other cases the overhead was negligible.

4.2 Memory Overhead

This section discusses the memory overhead of our countermeasure. This is done by providing both an analytical description of our worst case scenario and providing a measurement of the memory overheads that the benchmarks incur.

Theoretical memory overhead: An analytical study of memory usage has been conducted in the case of our highest level of security. This is achieved when we want to prevent the execution of the smallest shellcode by changing at least one character every 24 bytes. If s is the lenght of the smallest shellcode, the js_Transform() function will change the value of a random character every $(s-k)$ bytes, $k = 1...(s-1)$. In a real life scenario $k = 1$ is sufficient to guarantee a lack of space for the smallest shellcode. If the length of the original string is n bytes, the number of positions to transform will be $i = \lceil \frac{n}{s} \rceil$. The array used to store the position and the original value of the transform character will be $2i$ bytes long.

Memory usage: a numerical example Given the following data:

```
----------------------------------------------------------
original string length:  n = 1 MB = 1.048.576 bytes
smallest shellcode length: s = 25 bytes
injected sequence length:  r = 1  byte
----------------------------------------------------------
```

The number of positions will be $i = \lceil \frac{1MB}{s} \rceil = 43691$ and the memory overhead for the worst case will be 8.3%.

Table 5. Memory overhead of BuBBle in action on three Javascript benchmarks suites

Benchmark	Mem. usage (MB)	BuBBle mem. usage (MB)	Mem. overh.
Sunspider	88	93	+5.6%
V8	219	229	+4.2%
Peacekeeper	148	157	+6.5%
Average			**+5.3%**

Memory overhead for the benchmarks: Table 5 contains measurements of the maximum memory in megabyte that the benchmarks used during their runtime. These values were measured by starting up the browser, running the benchmarks to completion and then examining the *VmHWM* entry in $/proc/ < pid > /status$. This entry contains the peak resident set size which is the maximum amount of RAM the program has used during its lifetime. Our tests were run with swap turned off, so this is equal to the actual maximum memory usage of the browser. These measurements show that the overhead is significantly less than the theoretical maximum overhead.

4.3   Security Evaluation

In this section we give a security evaluation of BuBBle. When running the Javascript snippet of Fig.3 we are able to spray the heap in all cases: spraying means allocating memory and this is not considered an action to be detected. However, when attempting to execute the contents of sprayed objects, by a memory corruption, the attack will fail. The instruction pointer landed within a sled will execute the byte instruction 0xCC at a random position. This 1-byte instruction will call the interrupt procedure and execution will be halted. The execution of the 0xCC sequence is sufficient to consider the system under attack and to detect an unexpected execution. In fact, a legal access would purge the string of the 0xCC sequence. A drawback of our countermeasure is that a heap-spraying attack can still be performed by using a language other than Javascript such as Java or C#. However, the design in itself gives a reasonably strong security guarantee against heap-spraying attacks to be implemented for other browser supported languages. Another way to store malicious objects to the heap of the browser would be by loading images or media directly from the Internet. But this would generate a considerable amount of traffic and loading time, making the attack clearly observable.[5]

5   Related Work

Several countermeasures have been designed and implemented to specifically protect against heap-based attacks. Others have been designed to prevent memory corruption in general. We provide an overview of some countermeasures against heap overflow attacks in Section 5.2. A description of some countermeasures specifically designed to protect against heap-spraying attacks in web browsers is provided in Section 5.1.

---

[5] Heap spraying by content download might generate a traffic of hundreds of MBs. We are confident that also a broadband internet access would make the attack observable.

5.1 Heap-Spraying Defences

Nozzle: Nozzle is the first countermeasure specifically designed against heap-spraying attacks to web browsers [23]. It uses emulation techniques to detect the presence of malicious objects. This is achieved by the analysis of the contents of any object allocated by the Web browser. The countermeasure is in fact implemented at memory allocator level. This has the benefit of protecting against a heap-spraying attack by any scripting language supported by the browser. Each block on the heap is disassembled and a control flow graph of the decoded instructions is built. A `NOP-shellcode` object may be easily detected by this approach because one basic block in the control flow graph will be reachable by several directions (other basic blocks). For each object on the heap a measure of the likelihood of landing within the same object is computed. This measure is called `attack surface area`. The surface area for the entire heap is given by the accumulation of the surface area of individual blocks. This metric reflects the overall heap *health*. This countermeasure is more compatible than DEP and would help to detect and report heap-spraying attacks by handling exceptions, without just crashing. This approach has although some limitations. Because Nozzle examines objects only at specific times, this may lead to the so called TOCTOU-vulnerability (Time-Of-Check-Time-Of-Use). This means that an attacker can allocate a benign object, wait for Nozzle to examine it, then change it to contain malicious content and trigger the attack. Moreover Nozzle examines only a subset of the heap, for performance reasons. But this approach will lead to a lower level of security. The performance overhead of Nozzle examining the whole heap is unacceptable. Another limitation of Nozzle is the assumption that a heap-spraying attack allocates a relatively small number of large objects. A design based on this assumption would not protect against a heap-spraying which allocates a large number of small objects which will have the same probability to succeed.

Shellcode detection: Another countermeasure specifically designed against heap-spraying attacks to web browsers is proposed by [12]. This countermeasure is based on the same assumptions that (1) a heap-spraying attack may be conducted by a special crafted HTML page instrumented with Javascript and (2) Javascript strings are the only way to allocate contiguous data on the heap. Thus all strings allocated by the Javascript interpreter are monitored and checked for the presence of shellcode. All checks have to be performed before a vulnerability can be abused to change the execution control flow of the application. If the system detects the presence of shellcode, the execution of the script is stopped. Shellcode detection is performed by *libemu*, a small library written in C that offers basic x86 emulation. Since *libemu* uses a number of heuristics to discriminate random instructions from actual shellcode, false positives may still occur. Moreover an optimized version of the countermeasure that achieves accurate detection with no false positives is affected by a significant performance penalty of 170%.

5.2 Alternative Countermeasures

Probabilistic countermeasures: Many countermeasures make use of randomness when protecting against attacks. Canary-based countermeasures [21,24] use a secret random number that is stored before an important memory location: if the random number has changed after some operations have been performed, then an attack has been detected. Memory-obfuscation countermeasures [6,10] encrypt (usually with XOR) important memory locations or other information using random numbers. Memory layout randomizers [5,7,34] randomize the layout of memory: by loading the stack and heap at random addresses and by placing random gaps between objects. Instruction set randomizers [3] encrypt the instructions while in memory and will decrypt them before execution. While these approaches are often efficient, they rely on keeping memory locations secret. However, programs that contain buffer overflows could also contain "buffer overreads" (e.g. a string which is copied via *strncpy* but not explicitly null-terminated could leak information) or other vulnerabilities like format string vulnerabilities, which allow attackers to print out memory locations. Such memory leaking vulnerabilities could allow attackers to bypass this type of countermeasure. Another drawback of these countermeasures is that, while they can be effective against remote attackers, they are easier to bypass locally, because attackers could attempt brute force attacks on the secrets.

DEP: Data Execution Prevention [29] is a countermeasure to prevent the execution of code in memory pages. It is implemented either in software or hardware, via the NX bit. With DEP enabled, pages will be marked non-executable and this will prevent the attacker from executing shellcode injected on the stack or the heap of the application. If an application attempts to execute code from a page marked by DEP, an access violation exception will be raised. This will lead to a crash, if not properly handled. Unfortunately several applications attempt to execute code from memory pages. The deployment of DEP is less straightforward due to compatibility issues raised by several programs [2].

Separation and replication of information: Countermeasures that rely on separation or replication of information will try to replicate valuable control-flow information [36,37] or will separate this information from regular data. This makes it harder for an attacker to overwrite this information using an overflow. Some countermeasures will simply copy the return address from the stack to a separate stack and will compare it to or replace the return addresses on the regular stack before returning from a function. These countermeasures are easily bypassed using indirect pointer overwriting where an attacker overwrites a different memory location instead of the return address by using a pointer on the stack. More advanced techniques try to separate all control-flow data (like return addresses and pointers) from regular data, making it harder for an attacker to use an overflow to overwrite this type of data. While these techniques can efficiently protect against buffer overflows that try to overwrite control-flow information, they do not protect against attacks where an attacker controls an integer that is used as an offset from a pointer, nor do they protect against non-control-data attacks.

Execution monitors: In this section we describe two countermeasures that monitor the execution of a program and prevent transferring control-flow which could be unsafe. Program shepherding [20] is a technique that monitors the execution of a program and will disallow control-flow transfers[6] that are not considered safe. An example of

---

[6] Such a control flow transfer occurs when e.g., a *call* or *ret* instruction is executed.

a use for shepherding is to enforce return instructions to only return to the instruction after the call site. The proposed implementation of this countermeasure is done using a runtime binary interpreter. As a result, the performance impact of this countermeasure is significant for some programs, but acceptable for others. Control-flow integrity [1] determines a program's control flow graph beforehand and ensures that the program adheres to it. It does this by assigning a unique ID to each possible control flow destination of a control flow transfer. Before transferring control flow to such a destination, the ID of the destination is compared to the expected ID, and if they are equal, the program proceeds as normal. This approach, while strong and in the same efficiency range as our approach, does not protect against non-control data attacks.

6 Conclusion

Heap-spraying attacks expect to have large parts of the heap which are homogenous. By introducing heterogeneity where attackers expect this homogeneity, we can make heap-based buffer overflows a lot harder. By modifying the way the strings are stored in memory in the Javascript engine, we can achieve an effective countermeasure that introduces this heterogeneity. This is done by inserting special values at random locations in the string, which will cause a breakpoint exception to occur if they are executed. If an attacker tries to perform a heap-spraying attack, his injected code will now have been interrupted at a random location with such a breakpoint exception, allowing the browser to detect and report that an attack has occurred. Benchmarks show that this countermeasure has a negligible overhead both in terms of performance and memory overhead.

References

1. Abadi, M., Budiu, M., Erlingsson, Ú., Ligatti, J.: Control-flow integrity. In: Proceedings of the 12th ACM Conference on Computer and Communications Security, Alexandria, Virginia, U.S.A., November 2005, pp. 340–353. ACM, New York (2005)
2. Anisimov, A.: Defeating microsoft windows xp sp2 heap protection and dep bypass, http://www.ptsecurity.com
3. Barrantes, E.G., Ackley, D.H., Forrest, S., Palmer, T.S., Stefanović, D., Zovi, D.D.: Randomized instruction set emulation to disrupt binary code injection attacks. In: Proceedings of the 10th ACM Conference on Computer and Communications Security (CCS2003), Washington, D.C., U.S.A., October 2003, pp. 281–289. ACM, New York (2003)
4. Berry-Bryne, S.: Firefox 3.5 heap spray exploit (2009), http://www.milw0rm.com/exploits/9181
5. Bhatkar, S., Duvarney, D.C., Sekar, R.: Address obfuscation: An efficient approach to combat a broad range of memory error exploits. In: Proceedings of the 12th USENIX Security Symposium, Washington, D.C., U.S.A., August 2003, pp. 105–120. USENIX Association (2003)
6. Bhatkar, S., Sekar, R.: Data space randomization. In: Zamboni, D. (ed.) DIMVA 2008. LNCS, vol. 5137, pp. 1–22. Springer, Heidelberg (2008)
7. Bhatkar, S., Sekar, R., DuVarney, D.C.: Efficient techniques for comprehensive protection from memory error exploits. In: 14th USENIX Security Symposium, Baltimore, MD, August 2005, USENIX Association (2005)

8. Blog, M.A.L.: New backdoor attacks using pdf documents (2009),
   `http://www.avertlabs.com/research/blog/index.php/2009/02/`
   `19/new-backdoor-attacks-using-pdf-documents/`
9. Futuremark Corporation. Peacekeeper The Browser Benchmark,
   `http://service.futuremark.com/peacekeeper/`
10. Cowan, C., Beattie, S., Johansen, J., Wagle, P.: PointGuard: protecting pointers from buffer
    overflow vulnerabilities. In: Proceedings of the 12th USENIX Security Symposium, Wash-
    ington, D.C., U.S.A., August 2003, pp. 91–104. USENIX Association (2003)
11. Daniel, M., Honoroff, J., Miller, C.: Engineering heap overflow exploits with javascript. In:
    WOOT 2008: Proceedings of the 2nd conference on USENIX Workshop on offensive tech-
    nologies, Berkeley, CA, USA, pp. 1–6. USENIX Association (2008)
12. Egele, M., Wurzinger, P., Kruegel, C., Kirda, E.: Defending browsers against drive-by down-
    loads: mitigating heap-spraying code injection attacks. In: Flegel, U., Bruschi, D. (eds.)
    DIMVA 2009. LNCS, vol. 5587, pp. 88–106. Springer, Heidelberg (2009)
13. Erlingsson, Ú.: Low-level software security: Attacks and defenses. Technical Report MSR-
    TR-2007-153, Microsoft Research (November 2007)
14. Etoh, H., Yoda, K.: Protecting from stack-smashing attacks. Technical report, IBM Research
    Divison, Tokyo Research Laboratory (June 2000)
15. Mozilla Foundation. Firefox 3.5b4 (2009), `http://developer.mozilla.org`
16. Google. V8 Benchmark Suite - version 5, `http://v8.googlecode.com`
17. Intel. Intel architecture software developer's manual. vol. 2: Instruction set reference (2002)
18. E. C. M. A. International. ECMA-262: ECMAScript Language Specification. ECMA (Eu-
    ropean Association for Standardizing Information and Communication Systems), 3rd edn.,
    Geneva, Switzerland (December 1999)
19. Jorendorff: Anatomy of a javascript object (2008),
    `http://blog.mozilla.com/jorendorff/2008/11/17/`
    `anatomy-of-a-javascript-object`
20. Kiriansky, V., Bruening, D., Amarasinghe, S.: Secure execution via program shepherding. In:
    Proceedings of the 11th USENIX Security Symposium, San Francisco, California, U.S.A.,
    August 2002, USENIX Association (2002)
21. Krennmair, A.: ContraPolice: a libc extension for protecting applications from heap-
    smashing attacks (November 2003)
22. FireEye Malware Intelligence Lab. Heap spraying with actionscript (2009),
    `http://blog.fireeye.com/research/2009/07/`
    `actionscript_heap_spray.html`
23. Ratanaworabhan, P., Livshits, B., Zorn, B.: Nozzle: A defense against heap-spraying code
    injection attacks. Technical report, Microsoft Research (November 2008)
24. Robertson, W., Kruegel, C., Mutz, D., Valeur, F.: Run-time detection of heap-based over-
    flows. In: Proceedings of the 17th Large Installation Systems Administrators Conference,
    San Diego, California, U.S.A., October 2003, pp. 51–60. USENIX Association (2003)
25. securiteam.com. Heap spraying: Exploiting internet explorer vml 0-day xp sp2 (2009),
    `http://blogs.securiteam.com/index.php/archives/641`
26. Securitylab. Adobe reader 0-day critical vulnerability exploited in the wild, cve-2009-0658
    (2009), `http://en.securitylab.ru/nvd/368655.php`
27. skypher.com. Heap spraying (2007), `http://skypher.com/wiki/index.php`
28. Sotirov, A.: Heap feng shui in javascript (2007)
29. TMS. Data execution prevention,
    `http://technet.microsoft.com/en-us/library/cc738483.aspx`
30. Wagle, P., Cowan, C.: Stackguard: Simple stack smash protection for gcc. In: Proceedings of
    the GCC Developers Summit, Ottawa, Ontario, Canada, May 2003, pp. 243–256 (2003)

31. www2.webkit.org Sunspider javascript benchmark (2009),
    http://www2.webkit.org/perf/sunspider-0.9/sunspider.html
32. www.milw0rm.com Safari (arguments) array integer overflow poc (new heap spray)
    (2009), http://www.milw0rm.com/exploits/7673
33. www.packetstormsecurity.org 25bytes-execve (2009),
    http://www.packetstormsecurity.org/shellcode/
    25bytes-execve.txt
34. Xu, J., Kalbarczyk, Z., Iyer, R.K.: Transparent runtime randomization for security. In: 22nd
    International Symposium on Reliable Distributed Systems (SRDS 2003), Florence, Italy,
    October 2003, pp. 260–269. IEEE Computer Society, IEEE Press, Los Alamitos (2003)
35. Younan, Y., Joosen, W., Piessens, F.: Code injection in C and C++: A survey of vulnerabilities
    and countermeasures. Technical report, Departement Computerwetenschappen, Katholieke
    Universiteit Leuven (2004)
36. Younan, Y., Joosen, W., Piessens, F.: Efficient protection against heap-based buffer overflows
    without resorting to magic. In: Ning, P., Qing, S., Li, N. (eds.) ICICS 2006. LNCS, vol. 4307,
    pp. 379–398. Springer, Heidelberg (2006)
37. Younan, Y., Pozza, D., Piessens, F., Joosen, W.: Extended protection against stack smashing
    attacks without performance loss. In: Proceedings of the Twenty-Second Annual Computer
    Security Applications Conference (ACSAC 2006), pp. 429–438. IEEE Press, Los Alamitos
    (2006)

CsFire: Transparent Client-Side Mitigation of Malicious Cross-Domain Requests

Philippe De Ryck, Lieven Desmet, Thomas Heyman, Frank Piessens,
and Wouter Joosen

IBBT-DistriNet
Katholieke Universiteit Leuven
3001 Leuven, Belgium
{philippe.deryck,lieven.desmet}@cs.kuleuven.be

Abstract. Protecting users in the ubiquitous online world is becoming more and more important, as shown by web application security –
or the lack thereof – making the mainstream news. One of the more
harmful attacks is cross-site request forgery (CSRF), which allows an
attacker to make requests to certain web applications while imperson-
ating the user without their awareness. Existing client-side protection
mechanisms do not fully mitigate the problem or have a degrading effect
on the browsing experience of the user, especially with web 2.0 techniques such as AJAX, mashups and single sign-on. To fill this gap, this
paper makes three contributions: first, a thorough traffic analysis on real-
world traffic quantifies the amount of cross-domain traffic and identifies
its specific properties. Second, a client-side enforcement policy has been
constructed and a Firefox extension, named CsFire (CeaseFire), has been
implemented to autonomously mitigate CSRF attacks as precise as possible. Evaluation was done using specific CSRF scenarios, as well as in
real-life by a group of test users. Third, the granularity of the client-side
policy is improved even further by incorporating server-specific policy
refinements about intended cross-domain traffic.

1   Introduction

Cross-Site Request Forgery (CSRF) is a web application attack vector that can
be leveraged by an attacker to force an unwitting user's browser to perform
actions on a third party website, possibly reusing all cached authentication credentials of that user. In 2007, CSRF was listed as one of the most serious web
application vulnerability in the OWASP Top Ten [15]. In 2008, Zeller and Felten
documented a number of serious CSRF vulnerabilities in high-profile websites,
among which was a vulnerability on the home banking website of ING Direct [23].

One of the root causes of CSRF is the abuse of cached credentials in cross-
domain requests. A website can easily trigger new requests to web applications
in a different trust domain without any user intervention. This results in the
browser sending out cross-domain requests, while implicitly using credentials
cached in the browser (such as cookies, SSL certificates or login/password pairs).

F. Massacci, D. Wallach, and N. Zannone (Eds.): ESSoS 2010, LNCS 5965, pp. 18–34, 2010.
© Springer-Verlag Berlin Heidelberg 2010

From a server point of view, these implicitly authenticated requests are legitimate and are requested on behalf of the user. The user, however, is not aware that he sent out those requests, nor that he approved them.

Currently, a whole range of techniques exist to mitigate CSRF, either by protecting the server application or by protecting the end-user (e.g. via a browser extension or a client-side proxy). Unfortunately, the server-side protection mechanisms are not yet widely adopted. On the other side, most of the client-side solutions provide only limited protection or can not deal with complex web 2.0 applications, which use techniques such as AJAX, mashups or single sign-on (SSO). As a result, even the most cautious web user is unable to appropriately protect himself against CSRF, without compromising heavily on usability. Therefore, it is necessary to construct more robust client-side protection techniques against CSRF, capable of dealing with current and next-generation web applications.

This paper presents the following contributions. First, it describes the results of an extensive, real-world traffic analysis. This analysis shows how many of the requests a browser makes are cross-domain and what their properties are. Second, we define CsFire, an autonomous client-side protection policy, which is independent of user-input or server-provided information. This policy will determine which cross-domain traffic is considered harmful and will propose an appropriate action, such as blocking the request or removing the implicit authentication credentials from the request. The paper also discusses how this policy can be enforced within the Firefox browser. Finally, this autonomous policy can be extended by server-specific refinements, to achieve a more flexible and fine-grained enforcement.

This research builds upon previous work [13]. The main achievements with respect to the previous preliminary results are (1) a much finer-grained policy allowing a secure-by-default solution without service degradation, and (2) a thorough evaluation of the proposed prototype.

The remainder of this paper is structured as follows. Section 2 provides some background information on CSRF, explores the current state-of-the-art and defines the requirements of client-side mitigation solutions. Next, the detailed traffic analysis results are presented, along with the autonomous client policy, in Section 3. Section 4 describes the client-side solution in great detail. The refinements with server-specific policies and its implementation are discussed in Section 5. Our solution is evaluated extensively by means of test scenarios, along with a group of test users. The evaluation results can be found in Section 6. In Section 7, the presented work is related to alternative mitigation techniques and, finally, Section 8 summarizes the contributions of this paper.

2   Background

This section provides some background information on CSRF and available countermeasures. We will also define the requirements for client-side mitigation against CSRF.

2.1 Cross-Site Request Forgery (CSRF)

HTTP is a stateless client-server protocol [5], which uses certain methods to transfer data between web servers and browsers. The two most frequently used HTTP methods are GET and POST. Conforming to the HTTP specification, GET methods are used to fetch data from the server, whereas POST methods are used to update server state. In practice however, GET and POST methods are used interchangeably, and both can trigger server-side state changes.

Because of the stateless nature of HTTP, session management is built on top of HTTP. This is typically done by means of cookies, which are small amounts of session-specific data, created by the server application and stored in the browser. Alternative approaches such as URL rewriting or hidden form parameters exist as well [17]. This session-specific data is sent back with each request to that particular web application, without any user intervention.

Similarly, HTTP basic authentication attaches encoded credentials to each individual request to enable server-side authentication and authorization. For user convenience, the credentials are typically cached within the browser, and are only requested once for the lifespan of the browser session.

Other (albeit lesser used) authentication schemes include client side SSL and IP-based access control [9], of which the latter is mainly used on intranets. This paper will focus on the first two authentication mechanisms.

To perform a successful CSRF attack, a number of conditions need to be met:

1. The target website must use implicit authentication, such as through cookies or HTTP authentication, as mentioned above.
2. The targeted user must already have been authenticated to the target web application, and the user's browser must have cached the authentication credentials.
3. An attacker forces the user's browser to make an HTTP request to the target web application.

When all three conditions are met, the browser will automatically add the implicit authentication, making the request appear as a legitimate one to the target server.

In general, CSRF attacks happen cross-domain, where an attacker tricks the user into connecting to another site. Strictly speaking, CSRF can also occur within the same domain, e.g. due to a script injection vulnerability or when multiple users can host a web site within the same domain (universities, ISP's, ...). The focus of this paper lies on the cross-domain CSRF attacks.

2.2 Existing Countermeasures

A number of CSRF protection techniques exist, by either protecting the server application or by protecting the end-user. This section briefly discusses both.

Client-side countermeasures: The most widespread countermeasure is the Same Origin Policy (SOP) [22], implemented in most browsers. This policy limits access to DOM properties and methods to scripts from the same 'origin', where origin is usually defined as the triple <domain name, protocol, tcp port>[1]. This prevents a malicious script on the website `evil.com` from reading out session identifiers stored in a cookie from `homebanking.com`, for example. Unfortunately, the protection offered by the SOP is insufficient. Although the SOP prevents the requesting script from accessing the cookies or DOM properties of a page from another origin, it does not prevent an attacker from making *requests* to other origins. The attacker can still trigger new requests and use cached credentials, even though the SOP prevents the attacker from processing responses sent back from the server.

On top of SOP, client-side countermeasures exist to monitor and filter cross-domain requests. They typically operate as a client-side proxy [9] or as an extension in the browser [18,23]. These countermeasures monitor outgoing requests and incoming responses, and filter out implicit authentication or block cross-domain requests. Unfortunately, these client-side mitigation techniques suffer from various problems, as will be discussed in Section 7.

Server-side countermeasures: A number of server-side mitigation techniques exists as well, of which the most popular class is the use of secret tokens [4,10,16]. Each response from the server embeds a secret token into the web page (e.g. a hidden parameter in an HTML form). For each incoming request, the server verifies that the received token originated from the server, and is correctly bound to the user's session. Since the SOP prevents the attacker from reading responses from other origins, this is an effective server-side countermeasure.

Another class of server-side countermeasure relies on the HTTP `referer` header to determine the origin of the request. Unfortunately, quite often browsers or proxies block this header for privacy reasons [2]. Furthermore, an attacker can spoof HTTP headers, e.g. via Flash or request smuggling [11,12].

More intrusively, Barth et al. propose an additional HTTP header to indicate the origin of the request [2]. This header should not be removed by the browser or proxies, and is less privacy-intrusive than the referer. In addition to such an `origin` header, W3C proposes to add additional headers to the responses as well, to give server-side cross-domain information to the browser [19].

2.3   Requirements for Client-Side Protection

As stated before, the quality and applicability of the client-side countermeasures is still inadequate. The client-side mechanisms are necessarily generic, as they have to work for every web application, in every application domain. This usually makes them too coarse grained, often resulting in too permissive or too restrictive enforcement. In addition, most countermeasures do not handle current technologies well, such as JavaScript, AJAX and single sign-on (SSO).

---

[1] This definition of 'origin' will be used throughout the remainder of this paper.

Therefore, we propose the following requirements for a client-side CSRF solution in a contemporary web 2.0 context.

R1. The client-side protection should *not depend on user input*. Nowadays, a substantial fraction of web requests in an average browsing session is cross-domain (see Section 6). It is infeasible for the user to validate requests. Furthermore, users can not be expected to know which third parties a web application needs to function correctly. Therefore, a transparent operation is essential.

R2. The protection mechanism should be *usable in a web 2.0 context*, without noticeable service degradation. The solution should support the dynamic interaction behavior of today's applications (i.e., 'mashups'), and embrace current and future web technologies.

R3. *Secure by default*. The solution should have minimal false negatives using its default configuration.

3   Secure Cross-Domain Policy

The previous sections have made clear that even though there are effective countermeasures available, the user is left vulnerable due to the slow adoption of these countermeasures. In this section we will propose a policy defining exactly which cross-domain traffic is allowed and when protective measures need to be taken. This fine-grained policy, which is the main contribution of this paper, allows the users to protect themselves from malicious attackers and ill-secured websites.

To be able to determine an effective policy, information about cross-domain traffic is of crucial importance. By analyzing common cross-domain traffic, we can determine which request properties the policy enforcement mechanism can use to base its decisions on.

3.1   Properties of Cross-Domain Traffic

We have collected real-life traffic from about 15 volunteers over a time period of 2 weeks, resulting in a total of 750334 requests. The analysis of this traffic has revealed a number of properties that can be used to determine a secure cross-domain policy. We will discuss the traffic by pointing out how the requests are distributed in the total data set. We will examine the data through the strict SOP, which uses the triple <full domain name, protocol, tcp port>, as well as the relaxed SOP, which only considers the actual domain name (e.g. `example.com`. These detailed results are consistent with one million earlier recorded requests, as reported in [13].

A first overview, presented in Table 1, shows the distribution between the different request methods (GET, POST and other). Striking is that for the strict SOP, almost 43% of the requests are cross-domain. For the relaxed SOP, this is nearly 33%. The number of cross-domain requests is dominated by GET requests, with the POST requests having a minimal share.

As far as the other methods are concerned, we have discovered a very small number of HEAD and OPTIONS requests. Due to their insignificant amount, we do not focus on this type of requests. We do acknowledge that they need to be investigated in future research.

Table 1. Traffic statistics: overview of all traffic (for each row, the percentages of the first 3 columns add up to the percentage of the last column)

	GET	POST	Other	Total
Cross-domain requests	320529	1793	0	322322
(strict SOP)	(42.72%)	(0.24%)	(0.00%)	(42.96%)
Cross-domain requests	242084	1503	0	243587
(relaxed SOP)	(32.26%)	(0.20%)	(0.00%)	(32.46%)
All requests	722298	28025	11	750334
	(96.26%)	(3.74%)	(0.00%)	(100.00%)

Table 2 shows a detailed analysis of the GET requests. The columns show how much requests carry parameters (Params), how many requests are initiated by direct user interaction, such as clicking a link or submitting a form (User), how many requests contain a `cookie` header (Cookies) and how many requests carry HTTP authentication credentials (HTTP Auth). The results show that approximately 24% of the GET requests contain parameters. Furthermore, we can see that a very small amount of the GET-requests, especially of the cross-domain GET requests, originate from direct user interaction. The data also shows that cookies are a very popular authentication mechanism, whereas HTTP authentication is rarely used for cross-domain requests.

Table 2. Traffic statistics: overview of GET requests (for each row, the percentages in the columns are independent of each other and calculated against the total in the last column)

	Params	User	Cookies	HTTP Auth	Total
Cross-domain requests	82587	1734	116632	26	320529
(strict SOP)	(25.77%)	(0.54%)	(36.39%)	(0.08%)	(42.72%)
Cross-domain requests	58372	1100	59980	1	242084
(relaxed SOP)	(24.11%)	(0.45%)	(24.78%)	(0.00%)	(32.26%)
All GET requests	168509	7132	411056	651	722298
	(23.33%)	(0.99%)	(56.91%)	(0.89%)	(96.26%)

The analysis of the POST requests is summarized in Table 3, categorized in the same way as the GET requests, except for the presence of parameters. This data shows the same patterns as for the GET requests, albeit on a much smaller scale. We do see a larger percentage of cross-domain requests being initiated by the user, instead of being conducted automatically. Again, the HTTP authentication mechanism suffers in popularity, but cookies are used quite often.

Table 3. Traffic statistics: overview of POST requests (for each row, the percentages in the columns are independent of each other and calculated against the total in the last column)

	User	Cookies	HTTP Auth	Total
Cross-domain requests	158	1005	0	1793
(strict SOP)	(8.81%)	(56.05%)	(0.00%)	(0.24%)
Cross-domain requests	25	753	0	1503
(relaxed SOP)	(1.66%)	(50.10%)	(0.00%)	(0.20%)
All POST requests	930	23056	96	28025
	(3.32%)	(82.27%)	(1.99%)	(3.74%)

3.2   Defining a Policy

A policy blocking all cross-domain traffic is undoubtedly the most secure policy. However, from the traffic analysis we can easily conclude that this would lead to a severely degraded user experience, which conflicts with the second requirement of a good client-side solution. A policy that meets all three requirements will have to be more fine-grained and sophisticated. To achieve this goal, the policy can choose from three options for cross-domain requests: the two extremes are either allowing or blocking a cross-domain request. The road in the middle leads to stripping the request from authentication information, either in the form of cookies or HTTP authentication headers.

In order to preserve as much compatibility as possible, we have chosen to use the relaxed SOP to determine whether a request is cross-domain or not. This is comparable to JavaScript, where a relaxation of the origin is also allowed. We will now define the policy actions for each type of request and where possible, we will add further refinements. We will start by examining POST requests, followed by the GET requests. An overview of the policy is given in Table 4.

For a relaxed SOP, the traffic analysis shows that only 0.20 % of the cross-domain requests are POST requests, of which 1.66% is the result of direct user interaction. Therefore, we propose to strip all POST requests, even the manually submitted POST requests. This effectively protects the user from potentially dangerous UI redressing attacks [14], while having a minimal effect on the user experience.

Even though the HTTP protocol specification [5] states that GET requests should not have state-altering effects, we will ensure that CSRF attacks using GET requests are also prevented. This means that the policy for GET requests will have to be fine-grained. Since requests with parameters have a higher risk factor, we will define different rules for GET requests carrying parameters and GET requests without any parameters. The traffic analysis has shown that cross-domain GET requests with parameters are very common, which means that the attack vector is quite large too. Therefore, we propose to strip all GET requests with parameters from any authentication information. The GET requests without any parameters are less risky, since they are not likely to have a state-altering effect. Therefore, we have decided to allow GET requests with no parameters, if the request originates from user-interaction (e.g. clicking on a link). This helps preserving the unaltered user experience, because otherwise, when the user is logged in on a certain website, such as Facebook, and follows a link to `www.facebook.com` in for an example a search engine, the authentication information would be removed, which requires the user to re-authenticate. GET requests without any parameters that are not the result of direct user interaction will be stripped to cover all bases. If such a request would depend on authentication information, a re-authentication will be necessary.

Table 4. The secure default policy for cross-domain traffic

	Properties		**Decision**
GET	Parameters		STRIP
	No parameters	User initiated	ACCEPT
		Not User initiated	STRIP
POST	User initiated		STRIP
	Not User initiated		STRIP

4   Mitigating Malicious Cross-Domain Requests

In the previous section we have determined a policy to counter CSRF attacks. The policy will be enforced by a few specific components, each with their own responsibility. One of these components is the policy information point (PIP), where all the available information is collected. This information can be leveraged by the policy decision point (PDP) to make a decision about a certain request. This decision is used by the policy enforcement point (PEP), which will provide active protection for the user. We have implemented this policy as CsFire, an extension[2] for Mozilla Firefox that incorporates each of these components. The technical details of CsFire will now be discussed.

4.1   The Firefox Architecture

Mozilla Firefox, the second most popular browser, comes with an open and extensible architecture. This architecture is fully aimed at accommodating possible browser extensions. Extension development for Firefox is fairly simple and is done using provided XPCOM components [21]. Our Firefox extension has been developed using JavaScript and XPCOM components provided by Firefox itself.

To facilitate extensions wishing to influence the browsing experience, Firefox provides several possibilities to examine or modify the traffic. For our extension, the following four capabilities are extremely important:

– Influencing the user interface using XUL overlays
– Intercepting content-influencing actions by means of the `content-policy` event

---

[2] The extension can be downloaded from `https://distrinet.cs.kuleuven.be/software/CsFire/`.

- Intercepting HTTP requests before they are sent by observing the `http-on-modify-request` event (This is the point where the policy needs to be enforced).
- Intercepting HTTP responses before they are processed by observing the `http-on-examine-response` event

4.2 Policy Enforcement in Firefox

When a new HTTP request is received, the PEP needs to actively enforce the policy to prevent CSRF attacks. To determine how to handle the request, the PEP contacts the PDP which can either decide to *allow* the request, *block* it or *strip* it from any authentication information.

Enforcing an allow or block decision is straightforward: allowing a request requires no interaction, while blocking a request is simply done by signaling an error to Firefox. Upon receiving this error message, Firefox will abort the request. Stripping authentication information is less straightforward and consists of two parts: stripping cookies and stripping HTTP authentication credentials. How this can be done will be explained in the following paragraphs.

Firefox 3.5 has introduced a *private browsing mode*, which causes Firefox to switch to a mode where no cookies or HTTP authentication from the user's database are used. Private browsing mode stores no information about surfing sessions and uses a separate cookie and authentication store, which is deleted upon leaving private browsing mode. Unfortunately, we were not able to leverage this private browsing mode to strip authentication information from cross-domain requests, due to some difficulties. The major setback is the fact that Firefox makes an entire context switch when entering private browsing mode. This causes active tabs to be reloaded in this private mode, which essentially causes false origin information and influences all parallel surfing sessions.

Another approach is to manually remove the necessary HTTP headers when examining the HTTP request, before it is sent out. This technique is very effective on the `cookie` headers, but does not work for `authorization` headers. These headers are either added upon receiving a response code of 401 or 407 or appended automatically during an already authenticated session. In the former case, the headers are available upon examining the request and can be removed, but in the latter case, they are only added after the request has been examined. Obviously, this poses a problem, since the headers can not be easily removed.

Investigating this problem revealed that to implement the private browsing mode, the Firefox developers have added a new load flag[3], `LOAD_ANONYMOUS`, which prevents the addition of any `cookie` or `authorization` headers. If we set this flag when we are examining the HTTP request, we can prevent the addition of the `authorization` header. This is not the case for the `cookie` header, but as mentioned before, the `cookie` header, which at this point is already added to the request, can be easily removed.

---

[3] Load flags can be set from everywhere in the browser and are checked by the Firefox core during the construction of the request.

4.3   Considerations of Web 2.0

The difficulty of preventing CSRF can not necessarily be contributed to the nature of the attack, but more to the complex traffic patterns that are present in the modern web 2.0 context. Especially sites extensively using AJAX, single sign-on (SSO) mechanisms or mashup techniques, which combines content of multiple different websites, make it hard to distinguish intended user traffic from unintended or malicious traffic. Web 2.0 techniques such as AJAX and SSO can be dealt with appropriately, but mashups are extremely difficult to distinguish from CSRF attacks. Our solution has no degrading effect on websites using AJAX and SSO, but can be inadequate on mashup sites depending on implicit authentication to construct their content.

A SSO session typically uses multiple redirects to go from the site the user is visiting to an SSO service. During these redirects, authentication tokens are exchanged. When the original site receives a valid authentication token, the user is authenticated. Since all these redirects are usually cross-domain, no cookies or HTTP authentication headers can be used anyway, due to the SOP restrictions implemented in browsers. The authentication tokens are typically encoded in the redirection URLs. Our extension is able to deal with these multiple redirects and does not break SSO sessions.

5   Server Contributions to a More Fine-Grained Policy

The policy up until now was completely based on information available at the client-side. Unfortunately, such a policy fails to reflect intentions of web applications, where cross-domain traffic may be desired in certain occasions. To be able to obtain a more fine-grained policy, containing per site directives about cross-domain traffic, we have introduced an optional server policy. This server policy can tighten or weaken the default client policy. For instance, a server can prohibit any cross-domain traffic, even if authentication information is stripped, but can also allow intended cross-domain traffic from certain sites.

The technical implementation of server-side policies are fairly straightforward: the server defines a cross-domain policy in a pre-determined location, using a pre-determined syntax. The policy syntax, which is based on JSON, is expressed in the ABNF metasyntax language [3] and is available online[4]. The policy is retrieved and parsed by the browser extension at the client side. The policy is used by the PDP when decisions about cross-domain traffic need to be made.

The server policy has been made as expressive as possible, without requiring too many details to be filled out. The server policy can specify whether the strict SOP or the relaxed SOP needs to be used. Next to this, a list of intended cross-domain traffic can be specified. This intended traffic consists of a set of origins and a set of destinations, along with the policy actions to take. We have also provided the option to specify certain cookies that are allowed, instead of

---

[4] http://www.cs.kuleuven.be/~lieven/research/ESSoS2010/serverpolicy-abnf.txt

stripping all cookies. Finally, we also allow the use of the wild card character *, to specify rules for all hosts. An example policy can be found in Figure 1.

```
{"strictDomainEnforcement": true,
  "intendedCrossDomainInteraction": [
    {"blockHttpAuth": false,
     "blockCookies": false,
     "methods": ["*"],
     "cookieExceptions": [],
     "origins": [{
        "host": "www.ticket.com",
        "port": 443,
        "protocol": "https",
        "path": "/request.php" }],
     "destinations"= [{
        "port": 443,
        "protocol": "https",
        "path": "/confirm.php" }]},
    {"blockHttpAuth": true,
     "blockCookies": true,
     "methods": ["getNoParam"],
     "cookieExceptions": ["language"],
     "origins": [{ "host": "*" }]
    }]}
```

Fig. 1. An example server policy

Technically, the server policy is enforced as follows: when the PDP has to decide about a certain request, the request is checked against the target server policy. If a match is found, the decision specified by the policy will be found. If no match is found, the request is handled by the default client policy.

At the time, the composition of the server policy and the *secure by default* client policy to a unified policy is very rudimentary. This needs to be refined, such that the server can introduce policy refinements, without severely compromising the client-side policy. These refinements are left to future research.

6   Evaluation

CSRF attacks, as described earlier, are cross-domain requests abusing the cached authentication credentials of a user, to make state-altering changes to the target application. Knowing whether a request is state altering or not, is very application specific and very hard to detect at the client-side. The solution we proposed in the previous sections, examines all cross-domain traffic (intended and unintended) and limits the capabilities of such cross-domain requests, thus also prevents CSRF attacks. The extension is evaluated using a testbed of specially created test scenarios, to confirm that the capabilities of cross-domain requests are indeed limited as specified by the policy. A second part of the evaluation is done by a group of test users, that have used the extension during their normal, everyday surfing sessions. This part of the evaluation will confirm that even though the extension intercepts all cross-domain traffic – and not only the CSRF attacks –, the user experience is not affected by this fact. We conclude by presenting a few complex scenarios, that have been tested separately.

6.1   Extensive Evaluation Using the Scenario Testbed

To evaluate the effectiveness of CSRF prevention techniques, including our own solution, we have created a suite of test scenarios. These scenarios try to execute a CSRF attack in all different ways possible in the HTTP protocol, the HTML specification and the CSS markup language. The protocol and language specifications have been examined for cross-domain traffic possibilities. Each possible security risk was captured in a single scenario. Where applicable, a scenario has a version where the user initiates the request, as well as an automated version using JavaScript. For completeness, an additional JavaScript version using timeouts was created. In total, we have used a test suite of 59 scenarios. For requests originating from JavaScript, no scenarios are created, since these requests are typically blocked by the SOP.

Some highlights of these testing scenarios are redirects, either by the `Location` header, the `Refresh` header[5] or the `meta`-tag. For CSS, all attributes that use an URL as a value are possible CSRF attack vectors.

The extension has been evaluated against each of these scenarios. For every scenario, the CSRF attack was effectively prevented. Some scenarios conducted a hidden CSRF attack, in which case the user does not notice the attack being prevented. In case the attack is clearly visible, such as by opening a link in the current window, the user is presented with an authentication prompt for the targeted site. This is an altered user experience, but since an unwanted attack is prevented, this can not be considered a degradation of the surfing experience.

When discussing related work, these test scenarios will be used for the evaluation of other CSRF prevention solutions.

6.2   Real-Life Evaluation

A group of more than 50 test users, consisting of colleagues, students and members of the OWASP Chapter Belgium, has used CsFire with the policy as defined in this paper for over three months. The extension provides a feedback button, where users can easily enter a comment whenever they encounter unexpected effects. The results of a 1 month time slice are presented in Table 5. These numbers show that the extension has processed 1,561,389 requests, of which 27% was stripped of authentication credentials. The feedback logged by the users was limited to 3 messages, which was confirmed verbally after the testing period.

---

[5] Even though the `Refresh` header is not part of the official HTTP specification, it is supported by browsers.

Apart from the transparent evaluation by a group of test users, certain specific scenarios have been tested as well. This includes the use of single sign-on services such as Shibboleth and OpenID. The use of typical web 2.0 sites such as Facebook or iGoogle was also included in the evaluation.

Table 5. A 1 month time slice of evaluation data

Number of processed requests	1,561,389
Number of ACCEPT decisions	1,141,807
Number of BLOCK decisions	0
Number of STRIP decisions	419,582
Number of feedback messages	3

The only minor issue we detected, with the help of the feedback possibility of the extension, was with sites using multiple top-level domains. For instance, when Google performs authentication, a couple of redirects happen between several `google.com` domains and the `google.be` domain. This causes certain important session cookies to be stripped, which invalidates the newly authenticated session. This problem occurs for example with the calendar gadget of iGoogle, as well as the login form for *code.google.com*. This issue has not been registered on other Google services, such as Gmail, or any other websites.

These issues show why it is very difficult for an autonomous client-side policy to determine legitimate cross-domain traffic from malicious cross-domain traffic. This problem shows the importance of a server-side policy, which could relax the client-side policy in such a way that requests between `google.be` and `google.com` would be allowed.

A side-effect from the way Firefox handles its tabs becomes visible when using multiple tabs to access the same website. If a user is authenticated in tab one, a session cookie has probably been established. If the user now accesses this site in another tab, using a cross-domain request, the cookies will be stripped. This will cause the sessions in both tabs to be invalidated, which is a minor degrading experience. This behavior is very application-specific, since it depends on the way the application handles authentication and session management. This behavior has been experienced on LinkedIn, but does not appear on Facebook or Wikipedia. This problem can be mitigated with the integration of tab-isolation techniques. Such techniques are not yet available for Mozilla Firefox, but are in place in Google Chrome [7] and Microsoft Gazelle [20].

7   Related Work

In this section, we discuss CSRF protection mechanisms that where an inspiration to our solution: *RequestRodeo* and the *Adobe Flash cross-domain policy*. We also discuss two competing solutions: *BEAP* and *RequestPolicy*. Finally, we discuss *BEEP*, which proposes server-enforced policies, which can lead to future improvements of CSRF protection techniques.

RequestRodeo: The work of Johns and Winter aptly describes the issues with CSRF and a way to resolve these issues [9]. They propose *RequestRodeo*, a client-side proxy which protects the user from CSRF attacks. The proxy processes incoming responses and augments each URL with a unique token. These tokens are stored, along with the URL where they originated. Whenever the proxy receives an outgoing request, the token is stripped off and the origin is retrieved. If the origin does not match the destination of the request, the request is considered suspicious. Suspicious requests will be stripped of authentication credentials in the form of cookies or HTTP `authorization` headers. RequestRodeo also protects against IP address based attacks, by using an external proxy to check the global accessibility of web servers. If a server is not reachable from the outside world, it is considered to be an intranet server that requires additional protection. The user will have to explicitly confirm the validity of such internal cross-domain requests.

By stripping authentication credentials instead of blocking the request, RequestRodeo makes an important contribution, which lies at the basis of this work. Protecting against IP address based attacks is novel, and could also be added to our browser extension using the same approach. Johns and Winter do encounter some difficulties due to the use of a client-side proxy, which lacks context information. They also rely on a rewriting technique to include the unique token in each URL. These issues gain in importance in a web 2.0 world, where web pages are becoming more and more complex, which can be dealt with gracefully by means of a browser extension. Our solution is able to use full context information to determine which requests are authorized or initiated by a user.

Adobe Flash: By default, the Adobe flash player does not allow flash objects to access content retrieved from other websites. By means of a server-provided configurable policy, Adobe provides the option to relax the *secure by default* policy [1]. The target server can specify trusted origins, which have access to its resources, even with cross-domain requests. This technique was a source of inspiration for our own server-provided policies.

One unfortunate side-effect with the Adobe cross-domain policy and the example above is that a lot of sites have implemented an *allow all* policy [23]. To obtain a secure unified policy, smart composition of client and server-provided policies is crucial.

BEAP (AntiCSRF): Mao, Li and Molloy present a technique called *Browser-Enforced Authenticity Protection* [14]. Their solution resembles our solution, in a sense that they also aim to remove authentication credentials and have implemented a Firefox extension. Their policy to determine suspicious requests is quite flexible and based on the fact that GET requests should not be used for sensitive operations. As this may hold in theory, practice tells us otherwise, especially in the modern web 2.0 world. GET requests not carrying an `authorization` header are not considered sensitive, which leaves certain windows of attack open. BEAP addresses one of these issues, namely UI redressing, by using a *source-set* instead of a single source, the origin of the page. All the origins in the set, which are the origins of all ancestor frames, need to match the destination of the new request before it is allowed.

We have tested the provided Firefox extension against various test scenarios. The extension only works effectively against cross-domain POST requests, which is an expected consequence of the protection policy they propose. Unfortunately, the provided extension does not remove the `authorization` header and only seems to remove the `cookie` header. Our solution proposes a more realistic and more secure policy, and contributes technically by providing a clean way to actually remove an `authorization` header in Firefox.

RequestPolicy: Samuel has implemented a Firefox extension against CSRF attacks [18]. RequestPolicy is aimed at fully preventing CSRF attacks, which is realized by blocking cross-domain traffic, unless the sites are whitelisted. The criteria used to identify suspicious cross-domain traffic are user interaction and a relaxed SOP. Whenever a user is directly responsible for a cross-domain request, by clicking on a link or submitting a form, the request is allowed. Otherwise, traffic going to another domain is blocked. the extension allows a way to add whitelisted sites, such that traffic from `x.com` is allowed to retrieve content from `y.com`. By default, the extension proposes some whitelist entries, such as traffic from `facebook.com` to `fbcdn.com`.

When testing the RequestPolicy extension against our test scenarios, we found that almost all CSRF attacks are effectively blocked. Only the attacks which are caused by direct user interaction succeeded, which was expected. Unfortunately, when testing the extension by surfing on the internet, our experience was severely degraded. For starters, opening search results on Google stopped working. The cause for this issue is that clicking a search result actually retrieves a Google page[6], which uses JavaScript to redirect the user to the correct website. This JavaScript redirect is considered suspicious and therefore blocked. Apart from the Google issue, other sites suffered very noticeable effects. For instance, the popular site `slashdot.org` suffers major UI problems, since the stylesheet is loaded from a separate domain. These issues do not occur in our solution, since we only strip authentication information, instead of completely blocking cross-domain traffic.

BEEP: Jim, Swamy and Hicks propose *Browser-enforced Embedded Policies*, which uses a server-provided policy to defeat malicious JavaScript. They argue that the browser is the ideal point to prevent the execution of malicious scripts, since the browser has the entire script at execution time, even if it is obfuscated or comes from multiple sources. Their solution is to inject a protection script at the server, which will be run first by the browser and validates other scripts.

Such a server-provided but client-enforced technique is very similar to our solution, which is able to use server-provided policies and enforces security at the client-side. The solution proposed in BEEP can be an inspiration to work towards unified client-server CSRF protection mechanisms.

---

[6] When investigating this issue, not much information about this issue was found. We have noticed that this effect does not happen consistently, but have not found any logic behind this Google behavior.

8   Conclusion

We have shown the need for an autonomous client-side CSRF protection mechanism, especially since the already existing server-side protection mechanisms fail to get widely adopted and the available client-side solutions do not suffice. We have provided an answer to this requirement with our client-side browser extension which effectively protects the user from CSRF attacks. A predefined policy is used to determine which cross-domain requests need to be restricted, by either blocking the request or stripping the request from authentication credentials.

This work builds on preliminary results presented in an earlier paper, but presents much more detailed results in every section. We have conducted an extensive traffic analysis, which resulted in a number of request properties that can be used to determine the appropriate policy action for a cross-domain request. The predefined client-side policy uses these fine-grained criteria to achieve a policy that protects the user, without influencing the user experience in a negative way. In case cross-domain traffic is intended, which is not known by the client, servers can provide a cross-domain policy specifying which cross-domain requests are allowed. The browser extension merges both the client-side policy and the server-provided policy, to preserve the available protection mechanisms but also to offer as much flexibility as possible.

The policy and the enforcement mechanism have been thoroughly evaluated against CSRF attack scenarios, which cover all possibilities to mount a CSRF attack using HTTP, HTML or CSS properties. We have also collected results from a group of test users, which have actively used the extension during their normal surfing sessions. Finally, we have evaluated the extension against a few complex scenarios, such as sites using a single sign-on (SSO) mechanism or mashup techniques. Aside from one minor issue with sites spanning multiple top-level domains, no degrading effects where monitored, while all CSRF attack scenarios where successfully prevented. Even on mashup sites and sites using SSO mechanisms, no problems where detected.

The solution in this paper is not yet perfect and there is still room for improvement. Future research will focus on the refinement of the composition of a client-side policy and server-provided policies. The policies need to be extended to include other traffic besides GET and POST. Finally, the use of other authentication mechanisms, such as for instance SSL authentication, needs to be further investigated to prevent CSRF attacks abusing such credentials.

Acknowledgements

This research is partially funded by the Interuniversity Attraction Poles Programme Belgian State, Belgian Science Policy, IBBT and the Research Fund K.U. Leuven.

We would also like to thank our colleagues, students and members of the OWASP Chapter Belgium, who volunteered to collect traffic information and test CsFire.

References

1. Adobe. Adobe Flash Player 9 security (July 2008)
2. Barth, A., Jackson, C., Mitchell, J.C.: Robust defenses for Cross-Site Request Forgery. In: Proceedings of the 15th ACM Conference on Computer and Communications Security (CCS 2008), pp. 75–88 (2008)
3. Crocker, D., Overell, P.: Augmented BNF for syntax specifications: ABNF (2008), http://tools.ietf.org/html/rfc5234
4. Esposito, D.: Take advantage of ASP.NET built-in features to fend off web attacks (January 2005), http://msdn.microsoft.com/en-us/library/ms972969.aspx
5. Fielding, R., Gettys, J., Mogul, J., Frystyk, H., Masinter, L., Leach, P., Berners-Lee, T.: Hypertext Transfer Protocol – HTTP/1.1, rfc2616 (1999), http://tools.ietf.org/html/rfc2616
6. Gamma, E., Helm, R., Johnson, R., Vlissides, J.: Design Patterns. Addison-Wesley, Reading (1995)
7. Chromium Developer Documentation, http://dev.chromium.org/developers/design-documents/process-models
8. Jim, T., Swamy, N., Hicks, M.: Defeating script injection attacks with browser-enforced embedded policies. In: WWW 2007: Proceedings of the 16th international conference on World Wide Web (2007)
9. Johns, M., Winter, J.: RequestRodeo: Client side protection against session riding. In: Proceedings of the OWASP Europe 2006 Conference (2006)
10. Jovanovic, N., Kirda, E., Kruegel, C.: Preventing Cross Site Request Forgery attacks. In: IEEE International Conference on Security and Privacy in Communication Networks (SecureComm), Baltimore, MD, USA (August 2006)
11. Klein, A.: Forging HTTP request headers with Flash (July 2006), http://www.securityfocus.com/archive/1/441014
12. Linhart, C., Klein, A., Heled, R., Orrin, S.: HTTP request smuggling. Technical report, Watchfire (2005)
13. Maes, W., Heyman, T., Desmet, L., Joosen, W.: Browser protection against Cross-Site Request Forgery. In: Workshop on Secure Execution of Untrusted Code (SecuCode), Chicago, IL, USA (November 2009)
14. Mao, Z., Li, N., Molloy, I.: Defeating Cross-Site Request Forgery Attacks with Browser-Enforced Authenticity Protection. LNCS. Springer, Heidelberg (2001)
15. OWASP. The ten most critical web application security vulnerabilities
16. OWASP. CSRF Guard (October 2008), http://www.owasp.org/index.php/CSRF_Guard
17. Raghvendra, V.: Session tracking on the web. Internetworking 3(1) (2000)
18. Samuel, J.: Request Policy 0.5.8, http://www.requestpolicy.com
19. van Kesteren, A.: Cross-origin resource sharing (March 2009), http://www.w3.org/TR/2009/WD-cors-20090317/
20. Wang, H.J., Grier, C., Moshchuk, A., King, S.T., Choudhury, P., Venter, H.: The Multi-Principal OS Construction of the Gazelle Web Browser. Microsoft Research Technical Report, MSR-TR-2009-16 (2009)
21. XPCOM - MDC (2008), https://developer.mozilla.org/en/XPCOM
22. Zalewski, M.: Browser Security Handbook (2008), http://code.google.com/p/browsersec/wiki/Main
23. Zeller, W., Felten, E.W.: Cross-Site Request Forgeries: Exploitation and prevention. Technical report (October 2008), http://www.freedom-to-tinker.com/sites/default/files/csrf.pdf

Idea: Opcode-Sequence-Based Malware Detection

Igor Santos[1], Felix Brezo[1], Javier Nieves[1], Yoseba K. Penya[2], Borja Sanz[1], Carlos Laorden[1], and Pablo G. Bringas[1]

[1] S³Lab
[2] eNergy Lab
University of Deusto
Bilbao, Spain
{isantos,felix.brezo,javier.nieves,yoseba.penya,
borja.sanz,claorden,pablo.garcia.bringas}@deusto.es

Abstract. Malware is every malicious code that has the potential to harm any computer or network. The amount of malware is increasing faster every year and poses a serious security threat. Hence, malware detection has become a critical topic in computer security. Currently, signature-based detection is the most extended method within commercial antivirus. Although this method is still used on most popular commercial computer antivirus software, it can only achieve detection once the virus has already caused damage and it is registered. Therefore, it fails to detect new variations of known malware. In this paper, we propose a new method to detect variants of known malware families. This method is based on the frequency of appearance of opcode sequences. Furthermore, we describe a method to mine the relevance of each opcode and, thereby, weigh each opcode sequence frequency. We show that this method provides an effective way to detect variants of known malware families.

Keywords: malware detection, computer security, machine learning.

1 Introduction

Malware (or malicious software) is every computer software that has harmful intentions, such as viruses, Trojan horses, spyware or Internet worms. The amount, power and variety of malware increases every year as well as its ability to avoid all kind of security barriers [1] due to, among other reasons, the growth of Internet.

Furthermore, malware writers use code obfuscation techniques to disguise an already known security threat from classic syntactic malware detectors. These facts have led to a situation in which malware writers develop new viruses and different ways for hiding their code, while researchers design new tools and strategies to detect them [2].

Generally, the classic method to detect malware relies on a signature database [3] (i.e. list of signatures). An example of a signature is a sequence of bytes that is

F. Massacci, D. Wallach, and N. Zannone (Eds.): ESSoS 2010, LNCS 5965, pp. 35–43, 2010.

always present in a concrete malware file and within the files already infected by that malware. In order to determine a file signature for a new malware executable and to finally find a proper solution for it, specialists have to wait until that new malware instance has damaged several computers or networks. In this way, malware is detected by comparing its bytes with that list of signatures. When a match is found the tested file will be identified as the malware instance it matches with. This approach has proved to be effective when the threats are known in beforehand, and it is the most extended solution within antivirus software.

Still, upon a new malware appearance and until the corresponding file signature is obtained, mutations (i.e. aforementioned obfuscated variants) of the original malware may be released in the meanwhile. Therefore, already mentioned classic signature-based malware detectors fail to detect those new variants [2].

Against this background we advance the state of art in two main ways. First, we address here a new method that is able to mine the relevance of an opcode (operational code) for detecting malicious behaviour. Specifically, we compute the frequency with which the opcode appears in a collection of malware and in a collection of benign software and, hereafter, we calculate a discrimination ratio based on statistics. In this way, we finally acquire a weight for each opcode. Second, we propose a new method to compute similarity between two executable files that relies on opcode sequence frequency. We weigh this opcode sequence frequency with the obtained opcode relevance to balance each sequence in the way how discriminant the composing opcodes are.

2   Mining Opcode Relevance

Opcodes (or operational codes) can act as a predictor for detecting obfuscated or metamorphic malware [4]. Some of the opcodes (i.e. mov or push), however, have a high frequency of appearance within malware and benign executables, therefore the resultant similarity degree (if based on opcode frequency) between two files can be somehow distorted. Hence, we propose a way to avoid this phenomenon and to give each opcode the relevance that it really has.

In this way, we have collected malware from the *VxHeavens* website [5] forming a malware dataset of 13189 malware executables. This dataset contains only PE executable files, and, more accurately, it is made up of different kind of malicious software (e.g. computer viruses, Trojan horses, spyware, etc). For the benign software dataset, we have collected 13000 executables from our computers. This benign dataset includes, for instance, word-processors, drawing tools, windows games, internet browsers, pdf viewers and so on.

We accomplish the following steps for computing the relevance of each opcode. First, we disassemble the executables. In this step, we have used *The NewBasic Assembler* [6] as the main tool for obtaining the assembly files. Second, using the generated assembly files, we have built an opcode profile file. Specifically, this file contains a list with the operational code and the un-normalized frequency within both datasets (i.e. benign software dataset and malicious software dataset). Finally, we compute the relevance of each opcode based on the frequency with which it appears in both datasets. To this extent, we use *Mutual Information* [7], $I(X;Y) = \sum_{y\epsilon Y} \sum_{x\epsilon X} p(x,y) \log\left(\frac{p(x,y)}{p(x)\cdot p(y)}\right)$. Mutual information is a measure that indicates how statistically dependant two variables are. In our particular case, we define the two variables as each opcode frequency and whether the instance is malware. In this way, X is the opcode frequency and Y is the class of the file (i.e. malware or benign software), $p(x,y)$ is the joint probability distribution function of X and Y, $p(x)$ and $p(y)$ are the marginal probability distribution functions of X and Y.

Furthermore, once we computed the mutual information between each opcode and the executable class (malware of benign software) and we sorted them, we created an opcode relevance file. Thereby, this list of opcode relevance can help us to achieve a more accurate detection of malware variations since we are able to weigh the similarity function using these calculated opcode relevance and reducing the noise that irrelevant opcodes can produce.

3 Malware Detection Method

In order to detect both malware variants we extract the *opcode sequences* and their frequency of appearance. More accurately, we define a program ρ as a sequence of instructions I where $\rho = (I_1, I_2, ..., I_{n-1}, I_n)$. An instruction is composed by an operational code (*opcode*) and a parameter or list of parameters. In this way, we assume that a program is made up of opcodes. These opcodes can gather into several blocks that we call *opcode sequences*.

More accurately, we assume a program ρ as a set of ordered opcodes o, $\rho = (o_1, o_2, o_3, o_4, ..., o_{n-1}, o_n)$, where n is the number of instructions I of the program ρ. A subgroup of opcodes is defined as an opcode sequence os where $os \subseteq \rho$, and it is made up of opcodes o, $os = (o_1, o_2, o_3, ..., o_{m-1}, o_m)$, where m is the length of the sequence of opcodes os.

First of all, we choose the length of *opcode sequences*. Afterwards, we compute the frequency of appearance of each opcode sequence. Specifically, we use *term frequency* [8], $tf_{i,j} = \frac{n_{i,j}}{\sum_k n_{k,j}}$, that is a weight widely used in information retrieval. More accurately, $n_{i,j}$ is the number of times the term $t_{i,j}$ (in our case opcode sequence) appears in a document d, and $\sum_k n_{k,j}$ is the total number of terms in the document d (in our case the total number of possible opcode sequences).

Further, we compute this measure for every possible opcode sequence of a fixed length n, acquiring by doing so, a vector $\vec{v}$ made up of frequencies of opcode sequences $S = (o_1, o_2, o_3, ..., o_{n-1}, o_n)$. We weigh the frequency of appearance of this opcode sequence using the weights described in section 2. To this extent, we define *weighted term frequency* (*wtf*) as the result of weighting the relevance of each opcode when calculating the *term frequency*. Specifically, we compute it as the result of multiplying *term frequency* by the calculated weight of every opcode in the sequence. In this way, $weight(o)$ is the calculated weight for the opcode o and $tf_{i,j}$ is the *term frequency measure* for the given opcode sequence, $wtf_{i,j} = tf_{i,j} \cdot \prod_{o\epsilon S} \frac{weight(o)}{100}$. Once we have calculated the *weighted term frequency*, we have the vector of *weighted opcode sequence frequencies.* $\vec{v} = (wtf_1, wtf_2, wtf_3, ..., wtf_{n-1}, wtf_n)$.

We have focused on detecting known malware variants in this method. In this way, what we want to provide is a similarity measure between two files. Once we extract the opcode sequences that will act as features, we have a proper representation of the files as two input vectors $\vec{v}$ and $\vec{u}$ of *opcode sequences*. Hereafter, we can calculate a similarity measure between those two vectors. In this way, we use *cosine similarity*, $sim(\vec{v}, \vec{u}) = \cos(\theta) = \frac{\vec{v} \cdot \vec{u}}{||\vec{v}|| \cdot ||\vec{u}||}$ [9]. Therefore, we think that this measure will give a high result when two versions of the same malware instance are compared.

4 Experimental Results

For the following experiment, we have used two different datasets for testing the system: a malware dataset and a benign software one. First, we downloaded a big malware collection from VxHeavens [5] website conformed by different malicious code such as trojan horses, virus or worms. Specifically, we have used the next malware families: Agobot, Bifrose, Kelvir, Netsky, Opanki and Protoride.

We have extracted the *opcode sequences* of a fixed length (n) with $n = 1$ and $n = 2$ for each malware and some of its variants. Moreover, we have followed the same procedure for the benign software dataset. Hereafter, we have computed the *cosine similarity* between each malware and its set of variants. Further, we have computed the similarity of the malware instance with the whole benign software dataset.

Specifically, we have performed this process for every malware executable file within the dataset. For each malware family, we have randomly chosen one of its variants as the known instance and we have computed the cosine similarity between this variant and the other variants of that specific malware family. Moreover, we have performed the same procedure with a set of benign software in order to test the appearance of false positives.

Fig. 1 shows the obtained results of the comparison of malware families and their variants for an opcode sequence length n of 1. In this way, nearly every malware variant achieved a similarity degree between 90% and 100%. Still, the results obtained when comparing with the benign dataset (see Fig. 2) show that the similarity degree is too high, thus, this opcode sequence length seems to be not appropriate.

For an opcode sequence length of 2, the obtained results in terms of malware variant detection(see Fig. 3) show that the similarity degrees are more distributed in frequencies, however, the majority of the variants achieved a relatively high results. In addition, Fig. 4 shows the obtained results for the benign dataset. In this way, the results are better for this opcode sequence length, being more frequent the low similarity ratios.

Summarizing, on one hand, for the obtained results in terms of similarity degree for malware variant detection, the most frequent similarity degree is in the 90-100% interval. Moreover, the similarity degree frequency decreases and so

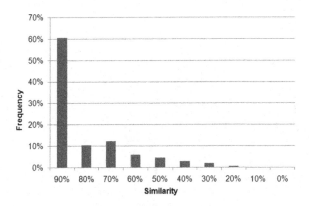

Fig. 1. A histogram showing the obtained results in terms of frequency of similarity ratio for the comparison of malware instances with their variants for n=1

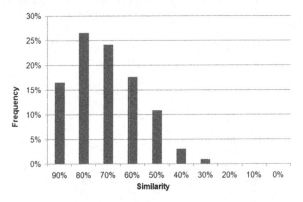

Fig. 2. A histogram showing the obtained results in terms of frequency of similarity ratio for the comparison of malware instances with benign executables for n=1

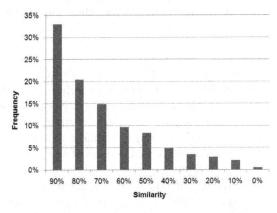

Fig. 3. A histogram showing the obtained results in terms of frequency of similarity ratio for the comparison of malware instances with their variants for n=2

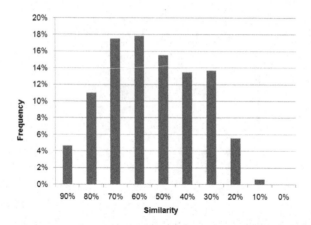

Fig. 4. A histogram showing the obtained results in terms of frequency of similarity ratio for the comparison of malware instances with benign executables for n=2

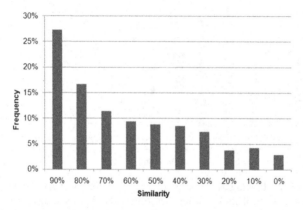

Fig. 5. A histogram showing the obtained results in terms of frequency of similarity ratio for the comparison of malware instances with their variants for n=1 and n=2 combined

the frequency does. Therefore, this method will be able to detect reliably a high number of malware variants after selecting the appropriate threshold of similarity ratio for declaring an executable as malware variant. Nevertheless, some of the executables were packed and, thereby, there are several malware variants that when computing the similarity degree did not achieve a high similarity degree.

Still, the similarity degrees between the two kind of sets (i.e. malware variants and benign software) are not different enough. Therefore, we decided to perform another experiment where the different opcode sequence lengths are combined ($n = 1$ and $n = 2$). Figures 5 and 6 show the obtained results. In this way, the malware variant similarity degrees remained quite high whilst the benign similarity degrees scrolled to lower results. On the other hand, for the obtained

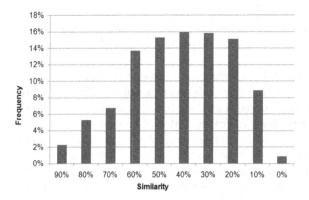

Fig. 6. A histogram showing the obtained results in terms of frequency of similarity ratio for the comparison of malware instances with benign executables for n=1 and n=2 combined

results in terms of similarity degree when comparing the malware instances with the benign dataset, as one may think in beforehand, the behaviour of the system yields to be nearly the opposite than when comparing it with its variants. In this way, the system achieved low similarity degrees in the majority of the cases of the benign dataset. Hence, if we select a threshold that allows us to detect the most number of malware variants as possible whilst the number of false positives is kept to 0, our method renders as a very useful tool to detect malware variants.

5 Related Work

There has been a great concern regarding malware detection in the last years. Generally, we can classify malware detection in *static* or *dynamic* approaches. Static detectors obtain features for further analysis without executing them since dynamic detectors execute malware in a contained environment.

In this way, static analysis for malware detection can be focused on the binary executables [10] or in source code [11] like the method proposed in this paper.

With regard to the binary analysis of the executables, there has been an hectic activity around the use of *machine-learning* techniques over byte-sequences. The first attempt of using non-overlapping sequence of bytes of a given length n as features to train a *machine-learning* classifier was proposed by Schulz et al. [12]. In that approach the authors proposed a method using the printable ASCII strings of the binary, tri-grams of bytes, the list of imported dynamically linked libraries (DLL), the list of DLL functions imported and the number of functions for each DLL. They applied multiple learning algorithms showing that multi-Naïve Bayes perform the best. Kolter et al. [13] improved the results obtained by Schulz et al. using n-grams (overlapping byte sequences) instead of non-overlapping sequences. Their method used several algorithms and the best results were achieved by a boosted decision tree. In a similar vein, a lot of work has been made over n-gram distributions of byte sequences and machine-learning [14]. Still, most of the features they used for the training of the classifiers can be changed easily by simply changing the compiler since they focus on byte distributions.

Moreover, several approaches have been based in the so-called *Control Flow Graph Analysis*. In this way, it is worth to mention the work of Christodescu and Jha [2] that proposed a method based of *Control Flow Analysis* to handle obfuscations in malicious software. Lately, Christodescu et. at. [15] improved the previous work including semantic-templates of malicious specifications. Nevertheless, the time resources they consume render them as not already full prepared to be adopted for antivirus vendors, although *Control Flow Analysis* techniques have proved to obtain some very valuable information of malicious behaviours.

Dynamic analysis for malware detection, as aforementioned, runs a program in a contained environment and collects information about it. Despite they are limited by one execution flow, they can overcome the main issue of static analysis: be sure that the code that will be executed is the one that is being analysed [16]. Therefore, these methods do not have to face obfuscations or in-memory mutation [17]. In this way, the safe environment can be based on a virtual machine [18] or based on DLL Injection and API Hooking [19].

6   Conclusions and Future Work

Malware detection has risen to become a topic of research and concern due to its increasing growth in past years. The classic signature methods that antivirus vendors have been using are no longer effective since the increasing number of new malware renders them unuseful. Therefore, this technique has to be complemented with more complex methods that provide detection of malware variants, in an effort of detecting more malware instances with a single *signature*.

In this paper, we proposed a method detecting malware variants that relied in opcodes sequences in order to construct a vector representation of the executables. In this way, based upon some length sequences, the system was able to detect the malicious behaviour of malware variants. Specifically, experiments have shown the following abilities of the system. First, the system was able to identify malware variants. Second, it was able to distinguish benign executables.

The future development of this malware detection system is oriented in three main directions. First, we will focus on facing packed executables using a hybrid dynamic-static approach. Second, we will expand the used features using even longer sequences and more information like system calls. Finally, we will perform experiments with a larger malware dataset.

References

1. Karsperky-Labs: Kaspersky Security Bulletin: Statistics 2008 (2009)
2. Christodorescu, M., Jha, S.: Static analysis of executables to detect malicious patterns. In: Proceedings of the 12th USENIX Security Symposium, February 2003, pp. 169–186 (2003)

3. Morley, P.: Processing virus collections. In: Proceedings of the 2001 Virus Bulletin Conference (VB 2001), Virus Bulletin, pp. 129–134 (2001)
4. Bilar, D.: Opcodes as predictor for malware. International Journal of Electronic Security and Digital Forensics 1(2), 156–168 (2007)
5. VX heavens (2009), http://vx.netlux.org/ (Last accessed: September 29, 2009)
6. NewBasic - An x86 Assembler/Disassembler for DOS,
   http://www.frontiernet.net/~fys/newbasic.htm
   (Last accessed: September 29, 2009)
7. Peng, H., Long, F., Ding, C.: Feature selection based on mutual information: criteria of max-dependency, max-relevance, and min-redundancy. IEEE Transactions on Pattern Analysis and Machine Intelligence, 1226–1238 (2005)
8. McGill, M., Salton, G.: Introduction to modern information retrieval. McGraw-Hill, New York (1983)
9. Tata, S., Patel, J.: Estimating the Selectivity of tf-idf based Cosine Similarity Predicates. SIGMOD Record 36(2), 75–80 (2007)
10. Carrera, E., Erdélyi, G.: Digital genome mapping–advanced binary malware analysis. In: Virus Bulletin Conference, pp. 187–197 (2004)
11. Ashcraft, K., Engler, D.: Using programmer-written compiler extensions to catch security holes. In: Proceedings of the 23rd IEEE Symposium on Security and Privacy, pp. 143–159 (2002)
12. Schultz, M., Eskin, E., Zadok, F., Stolfo, S.: Data mining methods for detection of new malicious executables. In: Proceedings of the 22nd IEEE Symposium on Security and Privacy, pp. 38–49 (2001)
13. Kolter, J.Z., Maloof, M.A.: Learning to detect malicious executables in the wild. In: Proceedings of the 10th ACM SIGKDD international conference on Knowledge discovery and data mining (KDD), pp. 470–478. ACM, New York (2004)
14. Santos, I., Penya, Y., Devesa, J., Bringas, P.: N-Grams-based file signatures for malware detection. In: Proceedings of the 11th International Conference on Enterprise Information Systems (ICEIS), Volume AIDSS, pp. 317–320 (2009)
15. Christodorescu, M., Jha, S., Seshia, S., Song, D., Bryant, R.: Semantics-aware malware detection. In: Proceedings of the 2005 IEEE Symposium on Security and Privacy, pp. 32–46 (2005)
16. Cavallaro, L., Saxena, P., Sekar, R.: On the limits of information flow techniques for malware analysis and containment. In: Zamboni, D. (ed.) DIMVA 2008. LNCS, vol. 5137, pp. 143–163. Springer, Heidelberg (2008)
17. Bayer, U., Moser, A., Kruegel, C., Kirda, E.: Dynamic analysis of malicious code. Journal in Computer Virology 2(1), 67–77 (2006)
18. King, S., Chen, P.: SubVirt: Implementing malware with virtual machines. In: 2006 IEEE Symposium on Security and Privacy, pp. 314–327 (2006)
19. Willems, C., Holz, T., Freiling, F.: Toward automated dynamic malware analysis using cwsandbox. IEEE Security & Privacy 5(2), 32–39 (2007)

Experiences with PDG-Based IFC

Christian Hammer

Purdue University[*]
cjhammer@purdue.edu

Abstract. Information flow control systems provide the guarantees that are required in today's security-relevant systems. While the literature has produced a wealth of techniques to ensure a given security policy, there is only a small number of implementations, and even these are mostly restricted to theoretical languages or a subset of an existing language.

Previously, we presented the theoretical foundations and algorithms for dependence-graph-based information flow control (IFC). As a complement, this paper presents the implementation and evaluation of our new approach, the first implementation of a dependence-graph based analysis that accepts full Java bytecode. It shows that the security policy can be annotated in a succinct manner; and the evaluation shows that the increased runtime of our analysis—a result of being flow-, context-, and object-sensitive—is mitigated by better analysis results and elevated practicability. Finally, we show that the scalability of our analysis is not limited by the sheer size of either the security lattice or the dependence graph that represents the program.

Keywords: software security, noninterference, program dependence graph, information flow control, evaluation.

1 Introduction

Information flow control is emerging as an integral component of end-to-end security inspections. There is very active research on how to secure the gap between access control and various kinds of I/O. Recent approaches to IFC are mainly based on special type sytems [20,21], and increasingly also on other forms of program analysis. This work reports on the experiences gained from our recent approach of the latter kind [14], and relates the insight gained with other tools based on type systems. The approach leveraged in this paper is based on *program dependence graphs* [8], more precisely *system dependence graphs* [15], which faithfully represent the semantics of a given program. The actual information flow validation is a graph traversal in the style of *program slicing*, which is a common program transformation on dependence graphs. The *(backward) slice* of a given statement contains all statements that may semantically influence that statement. Therefore, program slicing is closely connected to information

---

[*] This work was partially funded by Deutsche Forschungsgemeinschaft (DFG grant Sn11/9-2) at Karlsruhe University, Germany, where the experiments were conducted, and by the Office of Naval Research.

F. Massacci, D. Wallach, and N. Zannone (Eds.): ESSoS 2010, LNCS 5965, pp. 44–60, 2010.

flow control, which (in its simplest form, known as *noninterference*) checks that no secret input channel of a given program might be able to influence what is passed to a public output channel [11].

The major contributions of this work are:

- This paper presents the first dependence-graph-based IFC for full Java byte-code and its plugin integration into the Eclipse framework.
- Our IFC mechanism allows for succinct security policy specification as the number of required annotations is greatly reduced compared to traditional security type systems. This results a major improvement in practicability.
- The evaluation found that flow-, context-, and object-sensitive analysis pays off: While the analysis time increases compared to insensitive analyses, the results become significantly more precise, avoiding all sorts of false positives.
- Furthermore, the evaluation indicates that the analysis scales well to the security kernels we had in mind and that the scalability of our IFC analyses is not limited by either the size of the security lattice (which might be seen as a measure of the security policy's complexity), nor by the size of the program's dependence graph.

Previous work presented the theoretical foundations of dependence-graph-based IFC and described algorithms to efficiently implement these techniques. All these details are beyond the scope of this article, which focuses on presenting the framework for IFC in Eclipse and evaluating the effectiveness of these techniques. The interested reader is refered to our previous publications [12, 14] to find out all the details that cannot be covered in this work.

2   IFC Analysis

To illustrate dependence-graph-based IFC, Fig. 2 shows a (partial) SDG for Fig. 1. In the program, the variable sec is assumed to contain a secret value,

```
1  class A {
2    int x;
3    void set() { x = 0; }
4    void set(int i) { x = i;}
5    int get() { return x; }
6  }
7  class B extends A {
8    void set() { x = 1; }
9  }
10 class InfFlow {
11   PrintStream sysout = ...;
12   void main(String[] a){
13     //1. no information flow
14     int sec = 0 /*P:High*/;
15     int pub = 1 /*P:Low*/;
16     A o = new A();
17     o.set(sec);
18     o = new A();
19     o.set(pub);
20     sysout.println(o.get());
21     //2. dynamic dispatch
22     if (sec==0 && a[0].equals(
23       "007")) o = new B();
24     o.set();
25     sysout.println(o.get());
26   }
27 }
```

Fig. 1. An example program for information flow control

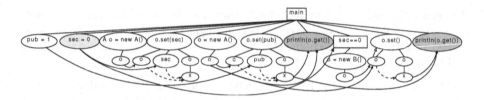

Fig. 2. SDG for the program in Figure 1

which must not influence printed output. First a new A object is created where field x is initialized to sec. However, this object is no longer used afterward as the variable is overwritten ("killed") with a new object whose x field is set to pub. Note that there is no path in the dependence graph from the initialization of sec to the first print statement, which means that sec cannot influence that output, as SDGs contain all possible information flows. This example demonstrates that in our SDG-based analysis the x fields in the two A objects are distinguished (object-sensitivity), the side-effects of different calls to set are not merged (context-sensitivity), and flow-sensitivity kills any influence from the sec variable to the first println. An analysis missing any of these three dimension of sensitivity would reject this part of the program, thus producing a false positive.

The next statements show an illegal flow of information: Line 22 checks whether sec is zero and creates an object of class B in this case. The invocation of o.set is dynamically dispatched: If the target object is an instance of A then x is set to zero; if it has type B, x receives the value one. Lines 22–24 are analogous to the following implicit flow:

$$\text{if (sec==0 \&\& ...) o.x = 0 else o.x = 1;}$$

In the PDG we have a path from sec to the predicate testing sec to o.set() and its target object o. Following the summary edge one reaches the x field and finally the second output node. Thus the PDG discovers that the printed value in line 25 depends on the value of sec.

3   Eclipse Plugins for IFC

We have implemented SDG-based IFC analysis, including declassification, as described in our previous work [14]. The prototype is an Eclipse plugin, which allows interactive definition of security lattices, automatic generation of SDGs, annotation of security levels to SDG nodes via source annotation and automatic security checks. At the time of this writing, all these components are fully operational.

We implemented the lattice editor based on Eclipse's GEF graph editing framework. The lattice elements are represented as bit vectors [1,9] to support fast infimum/supremum operators when checking for illegal information flow. It is worth noting that the algorithm of Ganguly et al. computes incorrect results without adding synthetic nodes into edges that span more than one level (where levels are defined in terms of the longest path between two nodes). The authors

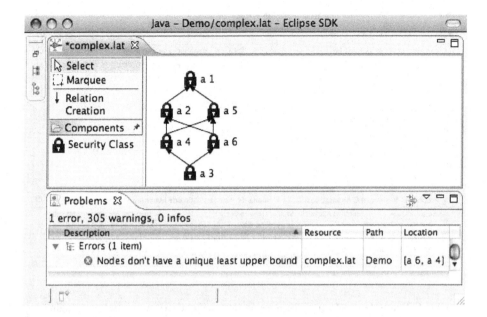

Fig. 3. The lattice editor in Eclipse with illegal graph

were not very specific concerning this restriction. Fortunately, these synthetic nodes can safely be removed from the lattice after conversion to a bit vector. Our editor attempts to convert the given graph to such a lattice. If this fails, the user is notified that the graph does not represent a valid lattice. Otherwise the lattice can be saved on disk for annotations. Figure 3 shows an example of a non-trivial graph with top element "a 1" and bottom element "a 3". But the lattice conversion fails for this graph, as the elements "a 4" and "a 6" do not have a unique upper bound: Both "a 2" and "a 5" are upper bounds for these elements, which violates a crucial property of lattices. The problem view at the bottom displays a detailed message to the user that describes the violation. If one of the relation edges between those four elements were removed, a valid lattice would be generated, which can be leveraged for annotating the source code.

The IFC algorithm was implemented in a two-stage version: As the first step, the so-called *summary declassification nodes* are computed. If the dependence graph already contained standard *summary edges* [15], these need to be removed first, as they would disturb computation of summary declassification nodes. Still, generating summary edges during SDG generation was not in vain: As summary declassification nodes can only arise between nodes that originally were connected by a summary edge, we can omit building the transitive closure of dependences for all nodes that are not the target of a summary edge. Algorithm 2 of [14] is therefore initialized with these actual-out nodes only. Note that this optimization does not improve the worst case complexity of the algorithm, but it reduces analysis time in practice (see section 5). As a second step, the IFC constraints are propagated through the backward slice of each output channel

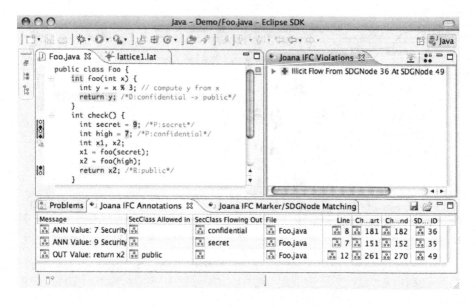

Fig. 4. Example program (with missing declassification) in our Eclipse plugin

according to Algorithm 1 of [14] . Our implementation does not generate these constraints explicitly, but performs a fixed point analysis.

Figure 4 shows an example program but is missing a declassification in the foo method. The Eclipse plugin features a full-fledged view for annotations and security violations. User annotations are shown in the Joana IFC Annotations view at the bottom of the figure. The message shows the kind of annotation (ANN stands for provided security level, OUT for required security level, and RED for declassification with both). Next to the message, the R and P annotations are shown. The rest of the entries describe the annotated source position and the ID of the SDG node. Another View, called "Joana IFC Marker/SDGNode Matching" allows precise matching of the selected source to its respective SDG node according to the debug information provided in the class file. The last view, depicted on the right in Figure 4 lists all violations found by the last IFC checking. For the example program, a run of the IFC algorithm determines a security violation between nodes 36 ($P(36) = secret$) and 49 ($R(49) = public$) because of the missing declassification in foo. When this declassification from *confidential* to *public* is introduced, no more illicit flow is detected.

4   Case Studies

As an initial micro-benchmark to compare our approach with type-based IFC, let us reconsider the program from Figure 1. Remember that PDG-based IFC guarantees that there is no flow from the secure variable (annotated $P(11) = High$) to the first output statement in line 20. Hence we analyzed the program from Figure 1 using Jif [18]. Jif uses a generalization of Denning's lattices, the so-called decentralized label model [19]. It allows to specify sets of security levels (called "labels" based on "principals") for every statement, and to attach a set of operations to any label. This is written e.g. $\{o_1 : r_1, r_2; o_2 : r_2; r_3\}$ and combines access control with IFC.

But note that the decentralized labels encode these access control features in an automatically generated standard security lattice and thus cannot overcome the imprecision of type-based analysis. As an example, we adapted the first part of Figure 1 to Jif syntax and annotated the declaration of o and both instantiations of A with the principal {pp:}. The output statement was replaced by an equivalent code that allowed public output. Jif reports that secure data could flow to that public channel and thus raised a false alarm. In fact, no annotation is possible that makes Jif accept the first part of Figure 1 without changing the source code.

4.1 A JavaCard Applet

As another case study for IFC we chose the JavaCard applet called `Wallet`.[1] It is only 252 lines long, but with the necessary API parts and stubs the PDG consists of 18858 nodes and 68259 edges. The time for PDG construction was 8 seconds plus 9 for summary edges on an Intel Core 2 Duo CPU with 2.33GHz and 2GB of RAM.

The Wallet stores a balance that is at the user's disposal. Access to this balance is only granted after supplying the correct PIN. We annotated all statements that update the balance with the provided security level *High* and inserted a declassification to *Low* into the `getBalance` method. The methods `credit` and `debit` may throw an exception if the maximum balance would be exceeded or if there is insufficient credit, resp. In such cases JavaCard applets throw an exception, and the exception is clearly dependent on the result of a condition involving balance. The exception is not meant to be caught but percolates to the JavaCard terminal, so we inserted declassifications for these exceptions as well. Besides this intended information flow, which is only possible upon user request and after verifying the PIN, our analysis proved that no further information flow is possible from the balance to the output of the JavaCard.

Note that this JavaCard applet—while operating on a restricted variant of Java—leverages many features of this standard: In particular we analyzed about 85 static variables and included all control flow due to implicit exceptions into our analysis without the need to explicitly declare or catch these. Some of these features would at least require reengineering of source code, others are explicitly prohibited by Jif, so this benchmark cannot be certified with Jif. Again we find that the increased precision of dependence graph based analysis allows a more permissive language.

4.2 The Battleship Example

The previous experiments demonstrated that our new approach is more general than Jif, because we can analyze realistic programming languages (in principle

---

[1] http://www.javaworld.com/javaworld/jw-07-1999/jw-07-javacard.html

Fig. 5. The lattice for analyzing the battleship example

all languages that compile to Java bytecode) and accept a larger number of secure programs due to increased precision. The next step in our evaluation will examine a native Jif program to get a direct comparison of practicability between these two systems. As a benchmark program, we chose the battleship example, which comes with every Jif installation and implements a non-GUI version of the popular battleship game. In this game, two players place ships of different lengths on a rectangular board and subsequently "bombard" random cells on the opponents board until one player has hit all the cells covered by adversary ships.

The source code of this program consists of about 500 lines plus the required libraries and stubs. These yield an SDG consisting of 10207 nodes and 77290 edges. For this example we use a standard diamond lattice, where $all \leq player \leq noOne$ and $all \leq other \leq noOne$ but neither $player \leq other$ nor $other \leq player$ (see Figure 5). This ensures that secret information of one player may not be seen by the other player and vice versa.

Before this example program could be analyzed by our IFC analysis, it had to be converted back to regular Java syntax. This included removal of all security types in the program, conversion of all syntactic anomalies like parentheses in throws clauses, and replacing all Jif peculiarities like its own runtime system. Most of this process required manual conversion. We annotated the ship placement strategy in the players initialization method with the security level $P(n) = player$. The three declassification statements of the original Jif program are modeled as declassifications from $player$ to all in our system as well. Then we annotated all parameters to `System.out.println` with $R(x) = all$, which corresponds to the original program's variable annotation.

When we checked the program with this security policy, illicit information flow was discovered to all output nodes. Manual inspection found that all these violations were due to implicit information flow from the players initialization methods, more precisely, due to possible exceptions thrown in these methods. However, closer inspection found that all of these exceptional control flow paths are in fact impossible.

As an example consider the initialization method in Figure 6. If the program checks whether a SSA variable is null, and only executes an instruction involving this variable if it is not (cf. line 25), then no null-pointer exception may ever be thrown at that instruction. However, our intermediate representation currently does not detect that fact, even if two identical checks for null are present in the intermediate representation, one directly succeeding the other. Jif supports such

```
1  /**
2   * Initialize the board by placing ships to cover numCovered coords.
3   */
4  void init/*{P:}**/(int/*{}*/numCovered) {
5  // Here what we would do in a full system is make a call to
6  // some non-Jif function, through the runtime interface, to
7  // get the position of the ships to place. That function would
8  // either return something random, or would implement some
9  // strategy. Here, we fake it with some fixed positions for
10 // ships.
11 final Ship/*[{P:}]**/[] myCunningStrategy = {
12      new Ship/*{P:}**/(new Coordinate/*[{P:}]**/(1, 1), 1, true),
13      new Ship/*{P:}**/(new Coordinate/*[{P:}]**/(1, 3), 2, false),
14      new Ship/*{P:}**/(new Coordinate/*[{P:}]**/(2, 2), 3, true),
15      new Ship/*{P:}**/(new Coordinate/*[{P:}]**/(3, 4), 4, false),
16      new Ship/*{P:}**/(new Coordinate/*[{P:}]**/(5, 6), 5, true),
17      new Ship/*[{P:}]**/(new Coordinate/*[{P:}]**/(5, 7), 6, false),
18 };
19
20 Board/*[{P:}]**/board = this.board;
21 int i = 0;
22      for (int count = numCovered; count > 0 && board != null; ) {
23      try {
24      Ship/*[{P:}]**/newPiece = myCunningStrategy[i++];
25      if (newPiece != null && newPiece.length > count) {
26          // this ship is too long!
27          newPiece = new Ship/*[{P:}]**/(newPiece.pos,
28                          count,
29                          newPiece.isHorizontal);
30      }
31      board.addShip(newPiece);
32      count -= (newPiece==null?0:newPiece.length);
33      }
34      catch (ArrayIndexOutOfBoundsException ignored) {}
35      catch (IllegalArgumentException ignored) {
36      // two ships overlapped. Just try adding the next ship
37      // instead.
38      }
39      }
40 }
```

Fig. 6. Initialization method of a Player in Battleship

local reasoning. For less trivial examples, where a final variable is defined in the constructor, and may thus never be null in any instance method, Jif requires additional annotation. We plan to integrate an analysis to detected such cases even in the interprocedural case [16]. Jif can only support non-local reasoning with additional user annotations.

Apart from null-pointer problems we found exceptional control flow due to array stores, where Java must ensure that the stored value is an instance of the array components, because of Java's covariant array anomaly. When a variable of an array type $a[]$ is assigned an array of a subtype $b[]$ where $b \le a$, then storing an object of type a into that variable throws an **ArrayStoreException**. Here Jif seems to have some local reasoning to prune trivial cases (see lines 11-17 in Figure 6). Our intermediate representation does currently not prune such cases, however, with the pointer analysis results we use for data dependences, such impossible flow could easily be removed.

Lastly, for interprocedural analysis, we found that our intermediate representation models exceptional return values for all methods, even if a method is guaranteed to not throw any exception. Pruning such cases can render the control flow in calling methods more precise and remove spurious implicit flow. Jif requires user annotations for such cases, as all possibly thrown exceptions must either be caught or declared, even `RuntimeExceptions`, which do not have to be declared in usual Java.

Currently, our tool does not offer such analysis, so there are only external means to detect such spurious cases: Either by manual inspection, theorem proving (e.g. pre-/post-conditions), or path conditions [13]. After verifying that such flow is impossible, we can block the corresponding paths in the dependence graphs, and we do that with declassification. In contrast to normal declassifications, where information flow is possible but necessary, this declassification models the guarantee that there is no information flow. As future work, we plan to integrate analyses which prune impossible control flow to reduce the false positive rate and thus the burden of external verification.

After blocking information flow through exceptions in Player's initialization, our IFC algorithm proved the battleship example secure with respect to the described security policy. No further illicit information flow was discovered. During the security analysis, based on only four declassifications, 728 summary declassification nodes were created. This result shows that summary declassification nodes essentially affect analysis precision, as they allow context-sensitive slicing while blocking transitive information flow at method invocation sites. Instead they introduce a declassification that summarizes the declassification effects of the invoked methods. Note that the original Jif program contained about 150 annotations (in Figure 6 these are shown as gray comments), most of which are concerned with security typing. Some of these annotations model, however, principals and their authority. Still the number of annotations is at least an order of magnitude higher than with our analysis. For a program that contains only 500 lines of code (including comments), this means that the annotation burden in Jif is considerable.

One of the reasons why slicing-based IFC needs less annotations than type systems is that side-effects of method calls are explicit in dependence graphs, so no end-label (which models the impact of side-effects on the program counter) is required, neither are return-value or exception labels. Those are computed as summary information representing the dependences of called methods.

Apart from these labels, Jif requires explicit annotations to verify any non-trivial property about exceptional control flow. In particular, many preconditions (e.g., non-nullness) need to be included into the program text instead of its annotations, e.g. explicit tests for null pointers or catch clauses, which are typically followed by empty handlers as in the example shown in Figure 6. Preconditions are therefore included as runtime tests to enable local reasoning. Such coding style is an ordeal from a software engineering perspective, as it impedes source comprehension and may conceal violated preconditions, which conflicts with Dijkstra's principle of the weakest precondition. What one really wants to have is verification that such cases cannot happen in any execution and thus do not need to be included into the source code.

5   Scalability

The previous sections demonstrated the precision and practicability of our approach. To validate the scalability of our new slicing-based information flow control, we measured execution times on a number of benchmarks with varying numbers of declassification and using lattices based on different characteristics. The benchmark programs are characterized in Table 1. We evaluated a benchmark of 8 student programs with an average size of 1kLoc, two medium-sized JavaCard applets and a Java application. The student programs use very few API calls, and for nearly all we designed stubs (for details see [12]) as to not miss essential dependences. The "Wallet" case study is the same as in section 4.1, the "Purse" applet is from the "Pacap" case study [5]. Both applet SDGs contain all the JavaCard API PDGs, native methods have been added as stubs. The program mp is the implementation of a mental poker protocol [3]. Again stubs have been used where necessary.

Table 2 shows the characteristics of the lattices we used in our evaluations: The first column displays the number of nodes in the lattice, the next column the maximal height of the lattice. The number of impure nodes in the lattice, which is shown in the next column, represents all nodes that have more than one parent in the lattice. The final column displays the number of bits needed in the efficient

Table 1. Data for benchmark programs

Nr	Name	LOC	Nodes	Edges	Time	Summary
1	Dijkstra	618	2281	4999	4	1
2	Enigma	922	2132	4740	5	1
3	Lindenmayer	490	2601	195552	5	10
4	Network Flow	960	1759	3440	6	1
5	Plane-Sweep	1188	14129	386507	24	13
6	Semithue	909	19976	595362	24	33
7	TSP	1383	6102	15430	15	2
8	Union Find	1542	13169	990069	36	103
9	JC Wallet	252	18858	68259	8	9
10	JC Purse	9835	135271	1002589	145	742
11	mp	4750	271745	2405312	141	247

Table 2. Characteristics of the lattices in the evaluation

	nodes	height	impure	bits
Lattice A	5	5	0	6
Lattice B	8	4	1	5
Lattice C	8	6	2	10
Lattice D	12	8	2	14
Lattice E	24	11	7	25
Lattice F	256	9	246	266

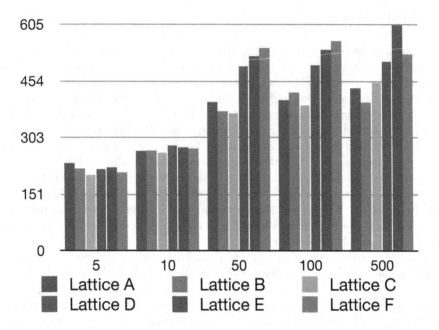

Fig. 7. Avg. execution time (y-axis, in s) of IFC analysis for the unionfind benchmark with different lattices and varying numbers of declassifications (x-axis)

bitset encoding of Ganguly et al. [9]. This encoding allows near-constant[2] computation of infima (greatest lower bounds), which will turn out to be essential for our evaluation. The lattices for evaluation have been designed such that they cover different characteristics equally: Lattice A is a traditional chain lattice, lattice B is more flat and contains an impure node. Lattice F has been automatically generated by randomly removing edges from a complete subset lattice of 9 elements. Conversion to bitset representation is only possible for the Hasse diagram, i.e. the transitive reduction partial order, which is not guaranteed by random removal of order edges. So we included a reduction phase before bitset conversion. Interestingly, Table 2 illustrates that the bitset conversion usually results in a representation with size linear in the number of lattice nodes.

Figure 7 shows the average execution time of 100 IFC analyses (y-axis, in seconds) for the unionfind benchmark of Table 1 using the lattices of Table 2. We chose the unionfind benchmark here, as it had the longest execution time, and the other benchmarks essentially show the same characteristics. For all IFC analyses we annotated the SDGs with 100 random security levels as provided and required security level, respectively. Moreover, we created 5 to 500 random declassifications to measure the effect of declassification on IFC checking (shown on the x-axis). The numbers illustrate that our IFC algorithm is quite independent of the lattice structure and size. In particular, we got a sub-linear increase

---

[2] The infimum computation is in fact constant, but we need hashing to map lattice elements to bitsets.

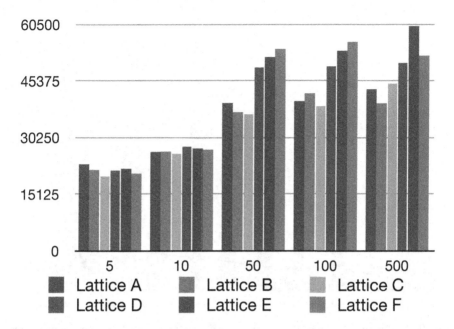

Fig. 8. Time for summary declassification nodes (in s) of unionfind with different lattices and varying numbers of declassifications (x-axis)

in execution time with respect to the lattice (and bitset) size. Apart from that, the increase with the number of declassifications is also clearly sub-linear, since the number of declassifications increases more than linear in our experiments (see y-axis). Figure 8 depicts the execution time for computing summary declassification nodes, which is a prerequisite for precise IFC checking (for details see [14, section 7]), therefore they have been acquired once for each combination of program, lattice, and declassifications. They were determined with the same random annotations as the numbers of Figure 7. Note that we did only compute summary information between nodes that were originally connected by summary edges. These numbers expose the same sub-linear correlations between time and lattice size or numbers of declassifications, respectively.

Figure 9 and 10 show the average execution time (y-axis, in seconds) of 100 IFC analyses and the time for summary declassification node computation, respectively, for all benchmark programs using the largest lattice and varying numbers of declassifications. Lines in this graph use the scale depicted on the right, while bars use a different scale, such that we included the numbers into each bar. For most programs, the analyses took less than a minute, with only semithue, purse, and unionfind requiring more time. Again, we found the correlation between execution time and number of declassifications sub-linear. In fact, the execution time for many benchmarks was lower with 500 declassifications than with 100. These numbers clearly illustrate the scalability of our information flow control analysis. There is no clear correlation between the number of nodes in the dependence graph and analysis time.

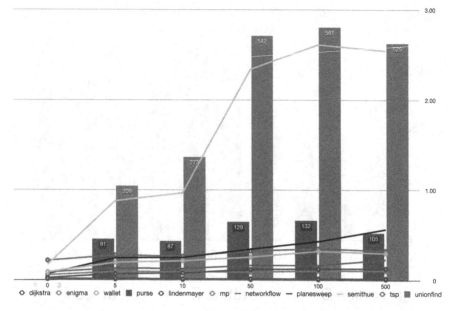

Fig. 9. Avg. execution time (y-axis, in s) of IFC analysis for all benchmark programs with the largest lattice and varying numbers of declassifications (x-axis). Bars use a different scale.

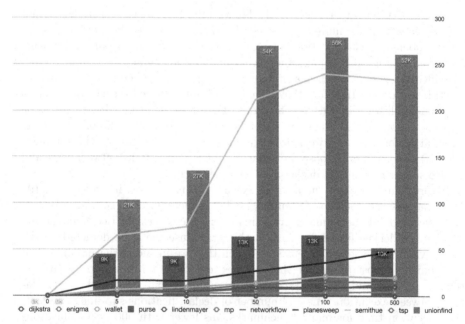

Fig. 10. Time for summary declassification nodes (y-axis, in s) for all benchmark programs with the largest lattice and varying numbers of declassifications (x-axis). Bars use a different scale.

However, there seems to be a correlation between the number of edges in the SDG and the execution time. Unlike slicing, our IFC analysis is not linear in the number of SDG nodes and edges, but must find a fixed point in the constraint system with respect to the given lattice. Therefore, it may traverse a cycle in the SDG as often as the lattice is high, and when cycles are nested this effect may become even worse. Our current implementation does not circumvent these effects, so one observes that the programs with the most edges yield to a substantial increase in analysis time. But note that the largest program, mp, does not belong to the outliers but behaves good-naturedly. One reason might be the different program structure, which can also be seen from original summary edge computation (see Table 1), which is considerably lower than for other large programs. This program does—unlike JavaCard applets and our student programs—not have a big loop in the main method which may invoke nearly all functionality. Concluding, we assume that the program's structure plays a bigger role than the pure number of nodes or edges for analysis time.

While future work must evaluate the impact of standard slicing optimizations on this technique for faster fixed point computation, we think that 1 minute execution time, as observed by the majority of our test cases, is definitely reasonable for a security analysis. But even the three outliers require maximally 1.5 hours (including summary declassification analysis), which should be acceptable for a compile-time analysis that usually needs to be done only once.

5.1 Future Work

While we presented evidence for precision, scalability, and practicability, there is still room for further improvements: In particular, we expect that optimizations for slicing, e.g., as presented by Binkley et al. [6], apply to our information flow analyses as well. These techniques produce up to 71% reduction in runtime and thus significantly improve scalability. Further research must evaluate which of these techniques are applicable to information flow control. Apart from that, compositionality is given for type systems but other analyses, like pointer analysis which is a prerequisite for our dependence graphs, are usually not. For more scalability of our approach, we plan to investigate how to make dependence-graph-based IFC compositional.

6 Related Work

Research in information flow control has been predominantly theoretic during the last decade (cf. e.g. [17, 20, 21]) where proposed systems were not implemented at all. More recently, increasingly more approaches are at least implemented for a rudimentary language, often just a while-language. With Java being a mainstream language, several approaches have targeted a core bytecode language (e.g. [2, 4]), but essential features like exceptions, constructors, or unstructured control flow are often not taken into account. Smith and Thober [22] present a type inference algorithm for a subset of Java which ameliorates the immoderate annotation burden of type based IFC. In contrast, the work described in this paper supports full Java bytecode with unstructured control flow, procedures, exception handling, etc., and shows that it scales to realistic security kernels.

Genaim and Spoto [10] define an abstract interpretation of the CFG looking for information leaks. It can handle all bytecode instructions of single-threaded Java and conservatively handles implicit exceptions of bytecode instructions. The analysis is flow- and context-sensitive but does not differentiate between fields of different objects. Instead, they propose an object-insensitive solution that folds together all fields of a given class. In our experience [12], object-insensitivity yields too many spurious dependences. The same is true for the approximation of the call graph by class hierarchy analysis. In this setting, both will result in many false alarms.

Chandra and Franz [7] implemented a hybrid IFC framework, where Java bytecode is analyzed statically and the security policy is checked dynamically, which allows dynamic updates of the policy. However, dynamic enforcement imposes a slowdown factor of 2. To improve performance of dynamic label computation, they only allow fully ordered sets instead of the general security lattices used in this work. It is, however, not clear if fast infimum computation [9] is actually slower than their more restrictive scheme. In contrast to their system we currently only allows a fixed security policy with purely static checking, which induces higher compile time overhead in favor of zero runtime overhead.

For several years, the most complete and elaborate system for Java-like languages has been Jif [18]. As noted before, Jif is neither an extension nor a restriction of Java, but an incompatible variant. Therefore it requires considerable reengineering efforts to convert a standard Java program to Jif. The newest version offers confidentiality as well as integrity checking in the decentralized label model. Our system differs from Jif in that it directly analyzes Java bytecode without requiring any refactoring of the source code. It, too, allows both integrity as well as confidentiality checking, though that entails manual definition of the security lattice, while Jif's decentralized label model automatically generates an appropriate lattice from the annotations. Still, our experiments indicate that the increased analysis cost is in fact mitigated by a significantly lower annotation burden and elevated analysis precision.

7   Conclusion

This paper evaluates our novel approach for information flow control based on system dependence graphs as defined in our previous work [14]. The flow-sensitivity, context-sensitivity, and object-sensitivity of our slicer extends naturally to information flow control and thus excels over the predominant approach for information flow control, which is security type systems.

The evaluation section showed that our new algorithm for information flow control dramatically reduced the annotation burden compared to type systems, due to its elevated precision. Furthermore, empirical evaluation showed the scalability of this approach. While it is clearly more expensive than security type systems, the evaluation demonstrates that security kernels are certified in reasonable time. As this certification process is only needed once at compile time, even an analysis that takes hours is acceptable when it guarantees security for the whole lifetime of a software artifact. As a consequence, this paper makes recent developments in program analysis applicable to realistic programming languages. The presented system implements the first dependence-graph-based information flow control analysis for a realistic language, namely Java bytecode. While dependence graph based IFC is not a panacea in that area, it nevertheless shows that program analysis has more to offer than just sophisticated type systems.

References

[1] Aït-Kaci, H., Boyer, R., Lincoln, P., Nasr, R.: Efficient implementation of lattice operations. ACM TOPLAS 11(1), 115–146 (1989)

[2] Amtoft, T., Bandhakavi, S., Banerjee, A.: A logic for information flow in object-oriented programs. In: POPL 2006, pp. 91–102. ACM, New York (2006)

[3] Askarov, A., Sabelfeld, A.: Security-typed languages for implementation of cryptographic protocols: A case study. In: di Vimercati, S.d.C., Syverson, P.F., Gollmann, D. (eds.) ESORICS 2005. LNCS, vol. 3679, pp. 197–221. Springer, Heidelberg (2005)

[4] Barthe, G., Pichardie, D., Rezk, T.: A certified lightweight non-interference Java bytecode verifier. In: De Nicola, R. (ed.) ESOP 2007. LNCS, vol. 4421, pp. 125–140. Springer, Heidelberg (2007)

[5] Bieber, P., Cazin, J., Marouani, A.E., Girard, P., Lanet, J.L., Wiels, V., Zanon, G.: The PACAP prototype: a tool for detecting Java Card illegal flow. In: Attali, I., Jensen, T. (eds.) JavaCard 2000. LNCS, vol. 2041, pp. 25–37. Springer, Heidelberg (2001)

[6] Binkley, D., Harman, M., Krinke, J.: Empirical study of optimization techniques for massive slicing. ACM TOPLAS 30(1), 3 (2007)

[7] Chandra, D., Franz, M.: Fine-grained information flow analysis and enforcement in a Java virtual machine. In: 23rd Annual Computer Security Applications Conference, pp. 463–475. IEEE, Los Alamitos (2007)

[8] Ferrante, J., Ottenstein, K.J., Warren, J.D.: The program dependence graph and its use in optimization. ACM TOPLAS 9(3), 319–349 (1987)

[9] Ganguly, D.D., Mohan, C.K., Ranka, S.: A space-and-time-efficient coding algorithm for lattice computations. IEEE Trans. on Knowl. and Data Eng. 6(5), 819–829 (1994)

[10] Genaim, S., Spoto, F.: Information flow analysis for Java bytecode. In: Cousot, R. (ed.) VMCAI 2005. LNCS, vol. 3385, pp. 346–362. Springer, Heidelberg (2005)

[11] Goguen, J.A., Meseguer, J.: Unwinding and inference control. In: Symposium on Security and Privacy, pp. 75–86. IEEE, Los Alamitos (1984)

[12] Hammer, C.: Information flow control for Java - a comprehensive approach based on path conditions in dependence graphs. Ph.D. thesis, Universität Karlsruhe (TH), Fak. f. Informatik (2009), URN urn=urn:nbn:de:0072-120494

[13] Hammer, C., Schaade, R., Snelting, G.: Static path conditions for Java. In: PLAS 2008, pp. 57–66. ACM, New York (2008)

[14] Hammer, C., Snelting, G.: Flow-sensitive, context-sensitive, and object-sensitive information flow control based on program dependence graphs. Int. Journal of Information Security 8(6), 399–422 (2009)

[15] Horwitz, S., Reps, T., Binkley, D.: Interprocedural slicing using dependence graphs. ACM TOPLAS 12(1), 26–60 (1990)

[16] Hubert, L.: A non-null annotation inferencer for Java bytecode. In: PASTE 2008, pp. 36–42. ACM, New York (2008)

[17] Hunt, S., Sands, D.: On flow-sensitive security types. In: POPL 2006, pp. 79–90. ACM, New York (2006)

[18] Myers, A.C., Chong, S., Nystrom, N., Zheng, L., Zdancewic, S.: Jif: Java information flow, http://www.cs.cornell.edu/jif/

[19] Myers, A.C., Liskov, B.: Protecting privacy using the decentralized label model. ACM TOSEM 9(4), 410–442 (2000)

[20] Sabelfeld, A., Myers, A.: Language-based information-flow security. IEEE Journal on Selected Areas in Communications 21(1), 5–19 (2003)

[21] Sabelfeld, A., Sands, D.: Declassification: Dimensions and principles. Journal of Computer Security 17(5), 517–548 (2009)

[22] Smith, S.F., Thober, M.: Improving usability of information flow security in Java. In: PLAS 2007, pp. 11–20. ACM, New York (2007)

Idea: Java vs. PHP: Security Implications of Language Choice for Web Applications

James Walden, Maureen Doyle, Robert Lenhof, and John Murray

Department of Computer Science
Northern Kentucky University
Highland Heights, KY 41099

Abstract. While Java and PHP are two of the most popular languages for open source web applications found at `freshmeat.net`, Java has had a much better security reputation than PHP. In this paper, we examine whether that reputation is deserved. We studied whether the variation in vulnerability density is greater between languages or between different applications written in a single language by comparing eleven open source web applications written in Java with fourteen such applications written in PHP. To compare the languages, we created a Common Vulnerability Metric (CVM), which is the count of four vulnerability types common to both languages. Common Vulnerability Density (CVD) is CVM normalized by code size. We measured CVD for two revisions of each project, one from 2006 and the other from 2008. CVD values were higher for the aggregate PHP code base than the Java code base, but PHP had a better rate of improvement, with a decline from 6.25 to 2.36 vulnerabilities/KLOC compared to 1.15 to 0.63 in Java. These changes arose from an increase in code size in both languages and a decrease in vulnerabilities in PHP. The variation between projects was greater than the variation between languages, ranging from 0.52 to 14.39 for Java and 0.03 to 121.36 in PHP for 2006. We used security and software metrics to examine the sources of difference between projects.

Keywords: web application security, security metrics, open source.

1 Introduction

While Java and PHP are two of the most popular languages for open source web applications found at Freshmeat [6], they have quite different security reputations. In this paper, we examine whether the variation in vulnerability density is greater between languages or between different applications written in a single language. We compare eleven open source web applications written in Java with fourteen such applications written in PHP. We also analyzed the source code of two different revisions of each of these applications to track the evolution of vulnerability density over time. PHP applications included both PHP 4 and PHP 5 code. The Java applications were compiled with Sun Java SE 6, but the 2006 versions of some applications had to be compiled with Sun Java SE 5.

F. Massacci, D. Wallach, and N. Zannone (Eds.): ESSoS 2010, LNCS 5965, pp. 61–69, 2010.
© Springer-Verlag Berlin Heidelberg 2010

Despite differences in security reputations, more than twice as many open source web applications are written in PHP than Java, and twelve of the fourteen PHP applications studied are more popular than any of the Java applications [6]. In part, PHP's poor security reputation [9] arises from default language features enabled in earlier versions of the language. However, these features have gradually been turned off as defaults or removed from the language. For example, the *register globals* feature which automatically created program variables from HTTP parameters was turned off as default in PHP 4.2 and removed in PHP 6.

We measured security through the number of vulnerabilities of types common to both languages as reported by a static analysis tool. Static analysis tools find common secure programming errors by evaluating source code without executing it. Static analysis has the advantage of being repeatable and checking all parts of the code equally, unlike human code reviewers or vulnerability researchers. The objective nature of static analysis makes it suitable for comparing different code bases, though, like human reviewers, static analysis tools make mistakes at times. We computed code size and complexity metrics and also a security resources indicator metric [15] to examine the source of differences between projects.

We discuss related work in section 2 and study design in section 3. Overall results are described in section 4, with section 5 analyzing results by vulnerability type. Sections 6 and 7 examine software and security metrics to determine the causes of differences between applications. Limitations of our analysis are discussed in section 8. Section 9 finishes the paper, giving conclusions and describing future work.

2   Related Work

Coverity used their Prevent static analysis tool to analyze a large number of open source projects written in C and C++ [3], using the static analysis vulnerability density metric. Fortify analyzed a small number of Java projects [5] with their static analysis tool, using the same metric. Nagappan used static analysis tools to measure defect density [8] to predict post-release defects. Note that defect density may not correlate with vulnerability density, as security flaws differ from reliability flaws.

Ozment and Schechter [12] and Li et. al. [7] studied how the number of security issues evolves over time. Ozment found a decrease in OpenBSD, while Li found an increase in both Mozilla and Apache.

Shin [14] and Nagappan et. al. [10] analyzed correlations of cyclomatic complexity with vulnerabilities. They had mixed results, with Shin finding a weak correlation for Mozilla and Nagappan finding three projects out of five having strong correlations. Shin also analyzed nesting complexity, finding significant but weak correlations with vulnerabilities for Mozilla.

Neuhaus and Zimmerman [11] studied the effect of dependencies on vulnerabilities in several thousand Red Hat Linux packages. Zimmerman et. al. [16] analyzed the problem of predicting defects based on information from other projects, finding that only 3.4% of cross-project predictions were both significant and had strong correlation coefficients.

3   Study Design

We examined the project history of 25 open source web applications, eleven of which were written in Java, fourteen of which were written in PHP. The applications are listed in table 1.

Table 1. Open Source Web Applications

Java			PHP		
alfresco	contelligent	daisywiki	achievo	obm	roundcube
dspace	jackrabbit	jamwiki	dotproject	phpbb	smarty
lenya	ofbiz	velocity	gallery2	phpmyadmin	squirrelmail
vexi	xwiki		mantisbt	phpwebsite	wordpress
			mediawiki	po	

To be selected, an application had to have a source code repository with revisions ranging from July 2006 to July 2008. The selected applications were the only applications that had revisions from those periods that could be built from source code in their repositories. While most third-party PHP libraries can be found in the PEAR or PECL repositories, third-party Java libraries are scattered among a variety of sites. Java developers often use tools like Maven to retrieve third-party software and manage builds.

Eight Java applications were not included in the study because they could not be built due to missing third-party software. Some older revisions used repositories of third-party tools that no longer existed, in which case we modified the Maven configuration to point to current repositories. This approach succeeded in some cases, but failed in others, as current Maven repositories do not contain every software version needed by older revisions. Some projects used other techniques to fetch dependencies, including ivy and custom build scripts.

Only five of the PHP projects and none of the Java projects maintained a public vulnerability database or had a security category in its bug tracker. While there were 494 Common Vulnerabilities and Exposures (CVE) listings for the PHP projects, there were only six such listings for the Java projects. The number of CVE entries does not necessarily indicate that a project is more or less secure. Due to the sparse and uneven nature of this data, documented vulnerabilities could not be used to measure the security of these applications. Instead, we used static analysis to find vulnerabilities in the source code of these applications.

We used Code Analyzer to compute SLOC, cyclomatic complexity, and nesting complexity for Java, and SLOCCount and Code Sniffer for PHP. We used Fortify Source Code Analyzer version 5.6 for static analysis. While there is no release quality free PHP static analysis tool, two of the Java web applications used the free FindBugs [1] static analysis tool. No web application showed evidence of use of a commercial static analysis tool in the form of files in the repository (which is how we identified use of FindBugs) or web site documentation.

Vulnerability density can be measured using the static analysis vulnerability density (SAVD) metric [15], which normalizes vulnerability counts by KSLOC (thousand source lines of code.) However, Fortify finds 30 types of vulnerabilities for Java and only 13 types for PHP in our set of applications, which prevents SAVD from being compared directly between the two languages. Since only four vulnerability types are shared between the two groups of applications we studied, we created a common vulnerability metric (CVM), which is the sum of those four vulnerability types, to more accurately compare results between Java and PHP. Common vulnerability density (CVD) is CVM normalized by KSLOC.

The four common vulnerability types were cross-site scripting, SQL injection, path manipulation, and code injection. Three of the four types are in the top five application vulnerabilities reported by MITRE in 2007 [2]. The two missing types from MITRE's top five are PHP remote file inclusion, which is found only in PHP, and buffer overflows, which are found in neither language.

4   Results

Examining the aggregate code base of the fourteen PHP applications, we found that common vulnerability density declined from 6.25 vulnerabilities/KSLOC in 2006 to 2.36 in 2008, a decrease of 62.24%. Over the same period, CVD declined from 1.15 in to 0.63 in the eleven Java applications, a decrease of 45.2%. Common vulnerabilities in PHP declined from 5425 to 3318, while common vulnerabilities increased from 5801 to 7045 in Java. The decrease in density for Java is the result of a tremendous increase in code size, from 5 million to 11 million SLOC. The expansion of the PHP code base was much smaller, from 870,000 to 1.4 million SLOC.

Java projects were larger on average than PHP projects. While one Java project, xwiki, had over a million of lines of code, the other ten Java projects ranged from 30,000 to 500,000 lines. The largest PHP project had 388,000 lines, and the smallest had under 6,000 lines, with the other twelve ranging from 25,000 to 150,000 lines. This difference tends to support the contention that PHP requires fewer lines of code to implement functionality than Java, especially as projects implementing the same type of software, such as wikis, were smaller in PHP than Java.

If we compare all vulnerability types, including all 30 categories of Java vulnerabilities and 13 categories of PHP vulnerabilities, we find that the vulnerability density of the Java code base decreased from 5.87 to 3.85, and PHP decreased from 8.86 to 6.02 from 2006 to 2008. The total number of PHP vulnerabilities increased from 7730 to 8459, while the total number of Java vulnerabilities increased from 29,473 to 42,581.

CVD varied much more between projects than between languages. In 2006, PHP projects ranged from 0.03 to 121.4 vulnerabilities/KLOC while Java projects had a much smaller range from 0.52 to 14.39. In 2008, both ranges shrank, with PHP projects varying from 0.03 to 60.41 and Java projects ranging from 0.04 to 5.96. Photo Organizer (po) had the highest CVD for both years. Figures 1 and 2 show change in vulnerability density between the initial and final revision for each project. In sections 6 and 7, we examine some possible sources of these differences between projects.

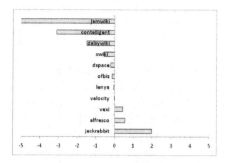

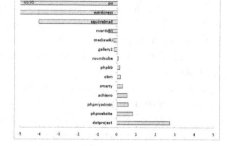

Fig. 1. Change in CVD for Java   **Fig. 2.** Change in CVD for PHP

5   Vulnerability Type Analysis

In this section, we examine the four vulnerability types that make up the CVM: cross-site scripting, SQL injection, path manipulation, and command injection. Figure 3 shows the changes in each vulnerability type between 2006 and 2008 for the aggregate Java and PHP code bases. The number of vulnerabilities in all four categories increased for Java, while they decreased for PHP.

Individual projects did not follow these overall patterns; two Java projects, contelligent and jamwiki, had reductions in three of the four categories. No Java project reduced the number of command injections. Two projects, alfresco and jackrabbit, did not reduce the number of vulnerabilities in any category.

Despite the overall decrease for PHP, nine of the fourteen PHP applications increased CVD. Two projects showed small decreases, while the remaining three contributed the bulk of the vulnerability reductions: photo organizer, squirrel-mail, and wordpress. Photo organizer is the only PHP project that saw a reduction in all four error types. Eight of the remaining PHP projects increased cross-site scripting errors, and nine increased path manipulation errors.

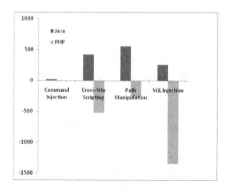

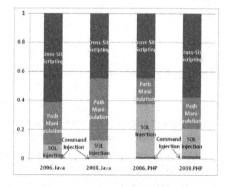

Fig. 3. Type Contribution to CVM   **Fig. 4.** Type Changes: 2006-2008

We also examined the contribution of each vulnerability type to the overall CVM and how that changed over the two years. Figure 4 compares the percentage contribution of each of the four vulnerabilities to the total CVM for Java and PHP projects in 2006 and 2008.

The 2008 ranking of the contributions of each error type for both languages and both years are the same: cross-site scripting, followed by path manipulation, SQL injection, and Command Injection. The total number of command injections is tiny compared to the other three types, which are found in MITRE's top five. The majority of the PHP change resulted from removing SQL injection flaws. Cross-site scripting vulnerabilities showed the largest decrease in Java, though the change was not as dramatic as the SQL injection reduction in PHP.

6   Software Metric Analysis

Based on prior work and research [3,8,10,12,13,15], we selected software metrics which had demonstrated correlations to vulnerability or defect density: cyclomatic complexity (CC) and nesting complexity. We used the same metric definitions as in [15], including three variants of each complexity metric: average, total, and maximum. Average is computed per-function for PHP and per-class for Java. While PHP 5 supports classes, these applications organized their code primarily with functions.

Figure 5 displays the correlations of metrics to CVD for both revisions. Correlation was computed using the Spearman rank correlation coefficient (ρ) since no assumptions can be made about the underlying distributions.

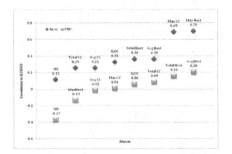
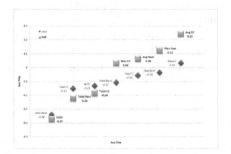

Fig. 5. ΔMetric correlations to ΔCVD   **Fig. 6.** Metric correlations to CVD

Significant correlations were found for maximum cyclomatic complexity and nesting complexity with change in CVD over the two year period ($p = 0.02$) for Java projects, but no correlations are signficant for the remaining metrics. While total code complexity is an indicator of changes in vulnerability density for Java projects, there are no significant correlations between software metrics and CVD for PHP projects.

We also compared change in metric values over the time period with change in CVD. We found only one signficant correlation; CVD is negatively correlated with SLOC for PHP projects. Since CVD decreased with time for this group of projects while SLOC increased, this result is not unexpected.

7 Security Resource Indicator

We measured the importance of security to a project by counting the public security resources made available on the project web site. We used the security resource indicator metric (SRI) [15], which is based on four items: security documentation for application installation and configuration, a dedicated e-mail alias to report security problems, a list of vulnerabilities specific to the application, and documentation of secure development practices, such as techniques to avoid common secure programming errors. The SRI metric is the sum of the four indicator items, ranging from zero to four.

Six of the eleven Java projects had security documentation, but none of the projects had any of the other three indicators. These results are similar to the results of Fortify's survey [5], in which only one of the eleven projects they examined had a security e-mail alias and two had links to security information. Their survey did not include the other components of the SRI metric.

PHP results were substantially different. While the percentage of projects with security documentation was lower, with only five of the fourteen projects having such documentation, six PHP projects had security e-mail contacts, five had vulnerability databases, and four had secure coding documentation. While there is no significant correlation of SRI with change in CVD, there is a significant correlation ($p < 0.05$) with a strong Spearman rank correlation coefficient, ρ, of 0.67, of SRI with change in SAVD, counting all PHP vulnerability categories.

The difference in SRI may result from the differences in application popularity. Open source PHP web applications are much more widely used than open source Java web applications. Popular projects are more likely to have vulnerabilities listed in the National Vulnerability Database [15], and therefore have a stronger incentive to track vulnerabilities and provide security contacts.

In addition to the greater number and higher Freshmeat popularity of PHP applications, language popularity is also revealed in what languages are supported by web hosting providers. Sixteen of the top 25 web hosting providers from `webhosting.info` listed supported languages: 87.5% supported PHP while only 25% supported Java. Several of the top hosting providers offered hosting for popular PHP applications, including Drupal, Joomla, Mambo, phpBB, and WordPress. None provided hosting for specific Java web applications.

8 Analysis Limitations

The 25 open source web applications were the only projects found on `freshmeat.net` that met our analysis criteria. Our analysis may not apply to other projects that were not analyzed in this work. Different static analysis tools look for different types of vulnerability and use different analysis techniques, so the vulnerability density from one tool cannot be compared directly to another. Static analysis tools also search for different vulnerabilities in different languages.

Static analysis tools report false positives, where a program mistakenly identifies a line of code as containing a vulnerability that it does not. Walden et al. [15] found that the Fortify static analysis tool had a false positive rate of 18.1% when examining web applications written in PHP. Coverity [3] found a false positive rate of less than 14% for millions of lines of C and C++ code.

9 Conclusion

We found that Java web applications had a substantially lower CVD than similar applications written in PHP, with 2008 values of 2.36 vulnerabilities/KSLOC for PHP and 0.63 for Java. Both sets of applications improved from 2006 to 2008, with PHP improving faster due to a decrease in vulnerability count while Java's improvement was due to a lower rate of vulnerabilities being inserted as code size grew. A large part of PHP's decrease was from a decline in SQL injection vulnerabilities, which could arise from higher usage of parameterized query interfaces as hosting providers offered newer versions of PHP database libraries.

The variation between projects was much greater than the variation between languages, ranging from 0.52 to 14.39 vulnerabilities/KSLOC for Java and 0.03 to 121.36 in PHP for 2006. Eight of the PHP projects had higher vulnerability densities in 2008 than 2006, while only three Java projects did. SRI was a useful predictor of how vulnerabilities evolved in PHP projects, but not for Java since none of the Java projects had security contacts or vulnerability listings. Complexity metrics were useful predictors for Java but not PHP vulnerability evolution.

In summary, programming language is not an important consideration in developing secure open source web applications. The correlation coefficient, $\rho = -0.07$, between language and CVD, was quite low, but it was not statistically significant. However, neither language had a clear advantage over the other in CVD over time and the variation between applications was much larger than the variation between languages.

References

1. Ayewah, N., Pugh, W.J., Morgenthaler, D., Penix, J.: Zhou. Y.: Evaluating Static Analysis Defect Warnings On Production Software. In: The 7th ACM SIGPLAN-SIGSOFT Workshop on Program Analysis for Software Tools and Engineering (June 2007)
2. Christey, S.M., Martin, R.A.: http://www.cve.mitre.org/docs/vuln-trends/index.html (published May 22, 2007)
3. Coverity, Coverity Scan Open Source Report 2009, http://www.coverity.com/scan/ (September 23, 2009)

4. Fenton, N.E., Pfleeger, S.L.: Software Metrics: A Rigorous and Practical Approach. Brooks/Cole, Massachusetts (1998)
5. Fortify Security Research Group and Larry Suto: Open Source Security Study (July 2008), http://www.fortify.com/landing/oss/oss_report.jsp
6. http://freshmeat.net/ (accessed September 27, 2009)
7. Li, Z., Tan, L., Wang, X., Lu, S., Zhou, Y., Zhai, C.: Have things changed now?: an empirical study of bug characteristics in modern open source software. In: Proceedings of the 1st workshop on Architectural and system support for improving software dependability, Association of Computing Machinery, New York, pp. 25–33 (2006)
8. Nagappan, N., Ball, T.: Static analysis tools as early indicators of pre-release defect density. In: Proceedings of the 27th International Conference on Software Engineering, Association of Computing Machinery, New York, pp. 580–586 (2005)
9. Shiflett, C.: PHP Security Consortium Redux, http://shiflett.org/blog/2005/feb/php-security-consortium-redux
10. Nagappan, N., Ball, T., Zeller, A.: Mining Metrics to Predict Component Failures. In: Proceedings of the 28th International Conference on Software Engineering, Association of Computing Machinery, New York, pp. 452–461 (2006)
11. Neuhaus, S., Zimmerman, T.: The Beauty and the Beast: Vulnerabilities in Red Hat's Packages. In: Proceedings of the 2009 USENIX Annual Technical Conference (USENIX 2009), San Diego, CA, USA (June 2009)
12. Ozment, A., Schechter, S.E.: Milk or Wine: Does Software Security Improve with Age? In: Proceedings of the 15th USENIX Security Symposium, USENIX Association, California, pp. 93–104 (2006)
13. Shin, Y., Williams, L.: An Empirical Model to Predict Security Vulnerabilities using Code Complexity Metrics. In: Proceedings of the 2nd International Symposium on Empirical Software Engineering and Measurement, Association for Computing Machinery, New York, pp. 315–317 (2008)
14. Shin, Y., Williams, L.: Is Complexity Really the Enemy of Software Security? In: Quality of Protection Workshop at the ACM Conference on Computers and Communications Security (CCS) 2008, Association for Computing Machinery, New York, pp. 47–50 (2008)
15. Walden, J., Doyle, M., Welch, G., Whelan, M.: Security of Open Source Web Applications. In: Proceedings of the International Workshop on Security Measurements and Metrics. IEEE, Los Alamitos (2009)
16. Zimmermann, T., Nagappan, N., Gall, H., Giger, E., Murphy, B.: Cross-project Defect Prediction. In: Proceedings of the 7th joint meeting of the European Software Engineering Conference and the ACM SIGSOFT Symposium on the Foundations of Software Engineering (ESEC/FSE 2009), Amsterdam, The Netherlands (August 2009)

Idea: Towards Architecture-Centric Security Analysis of Software

Karsten Sohr and Bernhard Berger

Technologie-Zentrum Informatik, Bremen, Germany
{sohr,berber}@tzi.de

Abstract. Static security analysis of software has made great progress over the last years. In particular, this applies to the detection of low-level security bugs such as buffer overflows, Cross-Site Scripting and SQL injection vulnerabilities. Complementarily to commercial static code review tools, we present an approach to the static security analysis which is based upon the software architecture using a reverse engineering tool suite called Bauhaus. This allows one to analyze software on a more abstract level, and a more focused analysis is possible, concentrating on software modules regarded as security-critical. In addition, certain security flaws can be detected at the architectural level such as the circumvention of APIs or incomplete enforcement of access control. We discuss our approach in the context of a business application and Android's Java-based middleware.

1 Introduction

More and more technologies find their way into our daily life such as PCs, mobile phones with PC-like functionality, and electronic passports. Enterprises, financial institutes, or government agencies map their (often security-critical) business processes to IT systems. With this dependency of technologies new risks go along which must be adequately addressed. One of the main problems arise from faulty software which led to most of the security incidents reported to CERT [3]. For this reason, a lot of work has been done on statically detecting common low-level security bugs in software such as buffer overflows, simple race conditions or SQL injection vulnerabilities. These efforts led to research prototypes [4,5,14] and even to commercial products [11,7,18]. In the future, however, it will be expected that attackers will also exploit other software errors such as logical flaws in access control (e.g., inconsistent role-based access control in Web Service applications) or the wrong usage of the security mechanisms provided by software frameworks such as JEE or Spring [17].

Moreover, even if the software architecture has been specified with modeling languages such as UML, it is not clear whether the actual software architecture manifesting in the source code is synchronized with the specified (intended) architecture. As software is often changed in an ad hoc fashion due to the customers demands, the software architecture is steadily eroded. In particular, this

F. Massacci, D. Wallach, and N. Zannone (Eds.): ESSoS 2010, LNCS 5965, pp. 70–78, 2010.

problem also applies to the security aspects of the software architecture. For example, if access to a security-critical resource ought to be protected by an API which enforces access control decisions, but an application for reasons of convenience can directly call the functionality in question without using this API, then the application's security can be easily subverted.

As a consequence, a methodology (including supporting tools) is desirable which allows one to analyze software on a more abstract level than the detailed source code. This way, certain kinds of logical security flaws in the software architecture can be detected as those mentioned above. In addition, a more abstract view on the software allows the security analyst to focus her analysis with already available tools more on the relevant software parts, reducing the rate of the false positives. For example, internal functionality, which is only called by a limited number of users, need not be analyzed to such an extent as modules exposed on the Internet. Last but not least, analyses at the architectural level can be carried out at design time such that security flaws can be detected at early stages of the software development process. This reduces the costs of resolving the security problems.

In this paper, we sketch a methodology for an architecture-based security analysis of software. The basis of this approach is the Bauhaus tool suite, a general-purpose reverse engineering tool, which has been successfully used in research projects as well as in commercial projects [19,21]. With the help of the Bauhaus tool, different kinds of analysis can be conducted at the architectural level. First, one can directly analyze the high-level architecture of an application w.r.t. security requirements. Due to the fact that the Bauhaus tool allows one to extract a low-level architecture from the source code, called resource flow graph (RFG), one can also check this architecture against the expected high-level architecture. This way, one can detect security problems such as missing access control checks in security-critical APIs.

Having carried out the analyses at this architectural level, one can switch to an analysis at the detail level, i.e., at the source code. This analysis can then be done with the help of commercially available tools or research tools such as model checkers and SAT solvers or Bauhaus itself.

The remainder of this paper is organized as follows. In Section 2, we describe the main concepts of the Bauhaus tool suite, concentrating on the RFG. We explain the principles of our architecture-based methodology for the security analysis of software in Section 3. Section 4 discusses some early results of our approach in the context of two different case studies, namely, a tutorial application for JEE and the Java-based middleware of the Android platform. Section 5 gives a short overview of related work, whereas Section 6 concludes and discusses possible research directions.

2   The Bauhaus Tool Suite

The Bauhaus tool suite is a reverse engineering tool which lets one deduce two abstractions from the source code, namely the Intermediate Language (IML) and the resource flow graph (RFG)[19]. The former representation in essence is an attributed syntax tree (an enhanced AST), which contains the detailed program information such as loop statements, variable definitions and name binding. The latter works on a higher abstraction level and represents architecturally relevant information of the software.

An RFG is a hierarchical graph, which consists of typed Nodes and edges. Nodes represent objects like routines, types, files and components. Relations between these elements are modeled with edges. The information stored in the RFG is structured in views. Each view represents a different aspect of the architecture, e.g., the call graph or the hierarchy of modules. Technically, a view is a subgraph of the RFG. The model of the RFG is fully dynamic and may be modified by the user, i.e., by inserting or deleting node/edge attributes and types. For visualizing the different views of RFGs, Graphical Visualiser (Gravis) has been implemented [8]. The Gravis tool facilitates high-level analysis of the system and provides rich functionality to produce new views by RFG analyses or to manipulate generated views.

3   Security Analyses with the Help of a RFG

We now discuss different aspects of analyzing software w.r.t. security based upon the RFG. Specifically, we describe how the methods and techniques, which have been well-established in the context of software quality assurance, can be adjusted for security analysis.

Notation of the architecture. Due to the fact that the elements of the RFG can be represented by a meta model, it is possible to define RFG profiles, that are specific to the software frameworks and security mechanisms used in the analyzed application. The elements of the profile have a well-defined semantics, which simplifies the analyses at the architectural level and leads to a better understanding of the security aspects of the architecture. For example, in case of a JEE business application one might have nodes of types such as "role", "permission", or "user" and edges of the types "user assignment" and "permission assignment" to express role-based access control (RBAC) [1]. In order to secure the communications, one might have "encrypted RPC channel" or "encrypted SOAP channel" edges.

Recovery of the software architecture. With the help of the reflexion method [15], the software architecture can be reconstructed and documented as a separate view of the RFG. This is done in a semi-automatic and iterative process starting from an abstract to a more detailed architecture description. Usually, this process is carried out in workshops with the software architects and developers.

Often the *mental* architectures, i.e., the architecture each developer / architect has in mind, might differ for the different participants of this process. These *mental* architectures are unified and written down as a hypothesized architecture which is then checked against the implemented low-level architecture. This way, discrepancies between both architectures, the hypothesized and the implemented one, can be automatically detected. If there are references (edges) in the implemented architecture which are absent in the hypothesized architecture, then we speak of *divergences*. An *absence* is a reference occurring in the hypothesized architecture and not being present in the source code. The architecture is manually refined and enhanced with new knowledge gained from previous steps in an iterative process until the architecture remains stable.

So far, the reflexion analysis was carried out having software quality in general rather than software security in mind. However, this step neatly fits to the "Ambiguity analysis" introduced by McGraw [17] because a different understanding of the software architecture might lead to security holes. This way, a natural application of the reflexion method in the security context is possible. In particular, the RFG representing the architecture can be enhanced with security modeling elements.

This reflexion method is used in 4.1 to find violations of the RBAC policy of the JEE demo system *Duke's Bank*.

Security views. As indicated in Section 2, the Bauhaus tool suite allows one to define views on the software architecture to concentrate on the aspects to be analyzed and hence to carry out a more focused analysis. In a JEE-based business application one might define a view which comprises all remote access or a view on RBAC. In the context of a mobile phone platform, one might define a view for the mechanism which implements the enforcement of permissions for protected resources such as Bluetooth or WiFi.

Further analyses on the RFG. Owing to the fact that applications process data of different sensitivity (security levels), the data flow through the modules of the software must be identified and the communications adequately secured. Those paths through an application's modules and functions should be identified where sensitive data flow without appropriate protection such as encryption or digital signatures. For this to accomplish, we can assign security labels to the data (e.g., member variables in Java or global variables in C) and also to the modules and functions of the application. We can further define which data can be accessed by which module and function, respectively. At the RFG level, we then can check whether the defined access control policy is violated.

Security requirements which the RFG must satisfy itself can be represented as a graph. An example of such a requirement is shown in Figure 1. Here, it is stated that if we have a remote method call on Entity Java Beans (EJBs), then the Java annotation "RolesAllowed" is mandatory. This means that remote method calls are only allowed if (the appropriate) roles are assigned to the caller's principal. One now can search for all matching occurrences of remote method calls on EJBs—the "condition" part in Figure 1—within the RFG and check whether the requirement is fulfilled for all occurrences of the condition. However, note that subgraph problems in general are known to be NP-complete such that heuristics are to be applied [12].

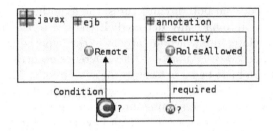

Fig. 1. A security requirement represented as a graph

Analyses on the RFG and the detailed program representation. As the Bauhaus tool suite also makes available the detailed program representation in form of the IML, code analyses can be carried out at the source code level. This way, one can start analyzing the software at the more abstract RFG level. Having identified security-relevant locations of the application via the RFG, one can analyze the corresponding source code locations more deeply, i.e., both representations can be used in conjunction for the security analysis. Alternatively, one can employ the RFG to pinpoint security-critical parts of the software and then use commercially available analysis tools to detect low-level security bugs. Research prototypes based on SAT solvers or theorem provers might also be useful in order to check the code against constraints (such as invariants or pre- and postconditions). Two promising examples of such tools are JForge [9] and ESC/Java2 [6].

4   Early Case Studies

We now discuss our architecture-centric security analysis in the context of two case studies. The first one is named "Duke's Bank", a simple application from Sun's JEE tutorial [20]. We chose this application because on the one hand, it is simple, and on the other hand, it has a widely-used architecture for business applications. In order to show that our approach can also be used in the context of embedded systems' software, the second case study is from the mobile phone domain, namely, the Java-based middleware of the Android platform.

4.1   Analysis of a JEE Application

Duke's Bank is a demo banking application allowing clerks to administer customer accounts and customers to access their account histories and perform transactions. It is a typical JEE application with a Web-based as well as a rich client interface (see Figure 2(a)). A customer can access information about his account via the Web interface, whereas the rich client interface can only be used by the clerk. The functionality of the application is provided by EJBs (`AccountControllerSessionBean`, `TxControllerSessionBean`, `CustomerControllerSessionBean`). In these EJBs, access to the database containing the account data is encapsulated via the Java Persistence framework.

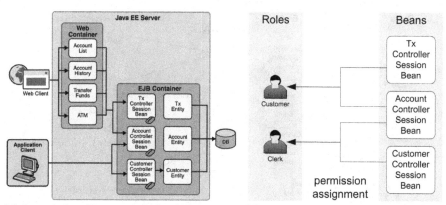

(a) Intended architecture of the *Duke's Bank* application [20]

(b) The extracted RBAC policy represented as a graph

Fig. 2. Description of the applications architecture

In Figure 2(a), one can see that the `CustomerControllerSessionBean` component cannot be accessed by the Web interface, i.e., the customers have no access to this bean. Figure 2(b) then displays the intended RBAC policy for the JEE application. The nodes represent the roles as well as EJBs and the edges correspond to the permission to access an EJB (permission assignment). Two roles *Clerk* and *Customer* have been defined, and specifically, there is no access from *Customer* to the component `CustomerControllerBean`.

We now briefly describe how the reflexion analysis can be applied to the *Duke's Bank* application with the focus on RBAC for EJBs. In a first step, we loaded the source code of this application into the Bauhaus tool and obtained an RFG as the low-level architecture, which is not given in this paper for reasons of brevity. The RBAC policy displayed in Figure 2(b) can then be regarded as the hypothesized architecture or more precisely as a security view representing the RBAC policy for the *Duke's Bank* application. This architecture is checked against the RFG gained from the source code.

Figure 3(a) shows the results of this reflexion analysis. Notice that the edges with the solid lines are representing divergences. The consequence of this divergence is that the code allows access that ought to be forbidden according to the RBAC policy, i.e., security violations are possible. The graph depicted in Figure 3(b) shows these violations manifesting in the source code's RFG [1]. For example, now every principal (be it a customer or a clerk) can access the method `getDetails()` of the `CustomerControllerSessionBean`. This way, she can query information on all customer data such as account numbers. Clearly, this is only a demo application, but it shows the kinds of problems which our analysis technique can detect.

---

[1] Note that the roles *Clerk* and *Customer* are mapped to the source code roles *bankAdmin* and *bankCustomer* within the frameworks of the reflexion analysis. Therefore, we have different names in Figure 3(b).

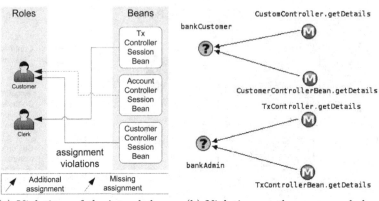

(a) Violations of the intended ar-   (b) Violations at the source code level
chitecture

Fig. 3. Analysis results

4.2   Analysis of Android

We also loaded the Android framework classes, implementing the Java-based middleware of Android into the Bauhaus tool suite to gain a better program understanding. Then we constructed an RFG, which can be used to identify the parts implementing the security concepts of the Android middleware such as mandatory access control for inter process communication (IPC), protected APIs, protection levels of applications [10]. This can be done by traversing the RFG and defining views which correspond to the security concepts implemented within the framework classes. The views let the analyst conduct a more focused analysis on the code. After constructing such views, she can switch to source code analysis as indicated above.

Employing our RFG-based analysis technique, we detected a kind of backdoor meant for Android developers, which does not seem to be officially documented: The end user can assign permissions to applications via a system-wide permission file. If there had not been additional security checks put in place, an end user could have given access rights defined by the operator to her own applications.

Beyond this, we can also check at the architectural level whether the permissions on security-critical resources (e.g., sending SMS, Bluetooth or WiFi access) are appropriately enforced. This means we can check on the RFG if the enforcePermission() methods are called for the permissions listed in the Manifest.permission class.

5   Related Work

There exist a plethora of works for the static security analysis of software. Some of those tools are research prototypes such as MOPS [4] and Eau Claire [5], others are successful commercial tools such as Fortify Source Code Analyzer [11],

Coverity Prevent [7], and Ounce [18]. Our approach is complementary to all those works because we utilize architectural information to focus the analysis on the source code and carry out our analyses *directly* on the architecture. Common low-level security bugs can clearly be detected by well-established analysis tools. Only, Coverity Prevent considers the software architecture for analyses, which at this time is limited to the software visualization not supporting more complex analyses such as the reflexion method.

In addition, as new modeling elements can be added to the RFG through meta-modeling, domain- and framework-specific security analyses can be performed. Little work has been done in the context of software frameworks before such as the LAPSE tool, which can find low-level security bugs in JEE applications [16].

Our work can also be compared with approaches to modeling and analyzing security requirements, specifically, in the context of UML profiles. Two such specification languages are UMLsec [13], which allows one to formulate security requirements w.r.t. access control and confidentiality, and SecureUML [2], which allows one to model RBAC policies. In addition, Basin et al. present an approach to analyzing RBAC policies based on UML meta-modelling [2]. Our work currently is focused on checking the intended architecture against the implemented low-level architecture, although analyzing the architecture itself remains future work.

6   Conclusion and Outlook

We presented an architecture-centric approach to the security analysis of software which can be seen as complementary work to available security review tools. This allows one to conduct the analyses on a more abstract level detecting also logical security flaws in the software such as erroneous RBAC of business applications. Our approach is based upon a reverse engineering tool suite called Bauhaus. With the help of two early case studies, we showed that our approach could be employed in different domains, namely, JEE-based business applications and mobile phones.

As this paper only reports on early results, a lot of work remains to be done in the future. First, we intend to systematically analyze more comprehensive business applications which employ software frameworks. Specifically, this will be done in the context of the Service Oriented Architecture. In addition, we intend to investigate how security views can be extracted and how far this process can be automated in the context of an encompassing case study such as the Android middleware. Last but not least, one can contemplate how to integrate our architecture-based analysis method into the Software Development Lifecycle with the focus on the continuous monitoring of the security architecture.

References

1. American National Standards Institute Inc. Role Based Access Control, ANSI-INCITS 359-2004 (2004)
2. Basin, D., Clavel, M., Doser, J., Egea, M.: Automated analysis of security-design models. Information and Software Technology 51, 815–831 (2009)

3. CERT/CC. CERT statistics (2008), http://www.cert.org/stats/
4. Chen, H., Wagner, D.: MOPS: an infrastructure for examining security properties of software. In: ACM Conference on Computer and Communications Security, pp. 235–244 (2002)
5. Chess, B.: Improving Computer Security Using Extended Static Checking. In: IEEE Symposium on Security and Privacy, p. 160 (2002)
6. Cok, D.R., Kiniry, J.: ESC/Java2: Uniting ESC/Java and JML. Technical report, University of Nijmegen (2004); NIII Technical Report NIII-R0413
7. Coverity. Coverity Prevent (2009), http://www.coverity.com
8. Czeranski, J., Eisenbarth, T., Kienle, H., Koschke, R., Simon, D.: Analyzing xfig Using the Bauhaus Tool. In: Working Conference on Reverse Engineering, pp. 197–199. IEEE Computer Society Press, Los Alamitos (2000)
9. Dennis, G., Yessenov, K., Jackson, D.: Bounded Verification of Voting Software. In: Shankar, N., Woodcock, J. (eds.) VSTTE 2008. LNCS, vol. 5295, pp. 130–145. Springer, Heidelberg (2008)
10. Enck, W., Ongtang, M., McDaniel, P.: Understanding Android Security. IEEE Security and Privacy 7(1), 50–57 (2009)
11. Fortify Software. Fortify Source Code Analyzer (2009), http://www.fortify.com/products/
12. Garey, M.R., Johnson, D.S.: Computers and Intractability. Freeman, San Francisco (1979)
13. Jürjens, J., Shabalin, P.: Automated verification of UMLsec models for security requirements. In: Baar, T., Strohmeier, A., Moreira, A., Mellor, S.J. (eds.) UML 2004. LNCS, vol. 3273, pp. 365–379. Springer, Heidelberg (2004)
14. Ashcraft, K., Engler, D.-R.: Using Programmer-Written Compiler Extensions to Catch Security Holes. In: IEEE Symposium on Security and Privacy, pp. 143–159 (2002)
15. Koschke, R., Simon, D.: Hierarchical Reflexion Models. In: Working Conference on Reverse Engineering, pp. 36–45. IEEE Computer Society Press, Los Alamitos (2003)
16. Livshits, V.B., Lam, M.S.: Finding Security Vulnerabilities in Java Applications Using Static Analysis. In: Proceedings of the 14th USENIX Security Symposium (August 2005)
17. McGraw, G.: Software Security: Building Security In. Addison-Wesley, Reading (2006)
18. Ounce Labs Inc. Website (2009), http://www.ouncelabs.com/
19. Raza, A., Vogel, G., Plödereder, E.: Bauhaus - A Tool Suite for Program Analysis and Reverse Engineering. In: Pinho, L.M., González Harbour, M. (eds.) Ada-Europe 2006. LNCS, vol. 4006, pp. 71–82. Springer, Heidelberg (2006)
20. Sun Microsystems. The Java EE 5 Tutorial (2008), http://java.sun.com/javaee/5/docs/tutorial/doc/bnclz.html
21. Universitaet Stuttgart. Project Bauhaus—Software Architecture, Software Reengineering, and Program Understanding (2009), http://www.bauhaus-stuttgart.de/bauhaus/index-english.html

Formally-Based Black-Box Monitoring of Security Protocols[*]

Alfredo Pironti[1] and Jan Jürjens[2]

[1] Politecnico di Torino
Turin, Italy
http://alfredo.pironti.eu/research
[2] TU Dortmund and Fraunhofer ISST
Dortmund, Germany
http://jurjens.de/jan

Abstract. In the challenge of ensuring the correct behaviour of legacy implementations of security protocols, a formally-based approach is presented to design and implement monitors that stop insecure protocol runs executed by such legacy implementations, without the need of their source code. We validate the approach at a case study about monitoring several SSL legacy implementations. Recently, a security bug has been found in the widely deployed OpenSSL client; our case study shows that our monitor correctly stops the protocol runs otherwise allowed by the faulty OpenSSL client. Moreover, our monitoring approach allowed us to detect a new flaw in another open source SSL client implementation.

1 Introduction

Despite being very concise, cryptographic protocols are quite difficult to get right, because of the concurrent nature of the distributed environment and the presence of an active, non-deterministic attacker. Increasing the confidence in the correctness of security protocol implementations is thus important for the dependability of software systems. In general exhaustive testing is infeasible, and for a motivated attacker one remaining vulnerability may be enough to successfully attack a system. In this paper, we focus in particular on assessing the correctness of legacy implementations, rather than on the development of correct new implementations. Indeed, it is often the case in practice that a legacy implementation is already in use which cannot be substituted by a new one: for example, when the legacy implementation is strictly coupled with the rest of the information system, making a switch very costly.

In this context, our proposed approach is based on black-box monitoring of legacy security protocols implementations. Using the Dolev-Yao [6] model, we assume cryptographic functions to be correct, and concentrate on their usage within the cryptographic protocols. Moreover, we concentrate on implementations of security protocol actors, rather than on the high level specifications of

---

[*] This research was partially supported by the EU project SecureChange (ICT-FET-231101).

F. Massacci, D. Wallach, and N. Zannone (Eds.): ESSoS 2010, LNCS 5965, pp. 79–95, 2010.

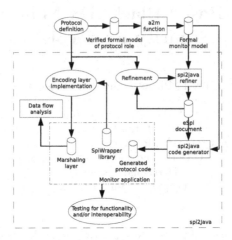

Fig. 1. Monitor design and development methodology

such security protocols. That is, we assume that a given protocol specification is secure (which can be proven using existing tools); instead, by monitoring it, we want to asses that a given implementation of one protocol's role is correct with respect to its specification, and it is resilient to Dolev-Yao attacks.

The overall methodology is depicted in figure 1. Given the protocol definition, a specification for one agent is manually derived. By using the "agent to monitor" ($a2m$) function introduced in this paper, a monitor specification for that protocol role is automatically generated. Then the monitor implementation is obtained by using the model driven development framework called spi2java [13], represented by the dashed box in the figure. The spi2java internals will be discussed later on in the paper. The monitor application is finally ran together with the monitored protocol role implementation (not shown in the picture).

A monitor implementation differs from a fresh implementation of a security protocol, because it does not execute protocol sessions on behalf of its users. The monitor instead observes protocol sessions started by the legacy implementations, in order to recognize and stop incorrect sessions, in circumstances where the legacy implementations cannot be replaced.

For performance trade-offs, monitoring can be performed either "online" or "offline". In the first case, all messages are first checked by the monitor, and then forwarded to the intended recipient only if they are safe. In the second case, all messages exchanged by the monitored application are logged, and then fed to the monitor for later inspection. The online paradigm prevents a security property to be violated, because protocol executions are stopped as soon as an unexpected message is detected by the monitor, before it reaches the intended recipient. However, online monitoring may introduce some latency. The offline paradigm does not introduce any latency and is still useful to recognize compromised protocol sessions later, which can limit the damage of an attack. For example, if a credit card number is stolen due to an e-commerce protocol attack, and if offline monitoring is run overnight, one can discover the issue at most one day later, thus limiting the time span of the fraud.

In this paper, the main goal of monitors is to detect, stop and report incorrect protocol runs. Monitors are not designed for example to assist one in forensic diagnosis after an attack has been found.

The monitoring is "black-box" in that the source code of the monitored application is not needed; only its observable behaviour (data transmitted over the medium, or *traces*) and locally accessed data are required. Thus any legacy implementation can still be used in production as is, while being monitored. The correctness of this approach depends on the correctness of the generated monitor. Our approach leverages formal methods in the derivation of the monitor implementation, so that a trustworthy monitor is obtained.

Note that this approach can be exploited during the testing phase as well: One can run an arbitrary number of simulated protocol sessions in a testing environment, and use the monitor to check for the correct behaviour.

In order to validate the proposed approach, a monitor for the SSL protocol is presented. The generated monitor stops incorrect sessions that could, for example, exploit a recently found flaw in the OpenSSL implementation.

The rest of the paper is organized as follows. Section 2 describes related work. Section 3 illustrates the formal background used in the paper. Section 4 describes the function translating a Spi Calculus protocol agent's specification into a monitor specification for that agent. Then section 5 shows the SSL protocol case study. Finally section 6 concludes.

For brevity, this paper mainly concentrates on the description of the proposed approach and on its validation by means of a real-life size case study. An extended version of this paper that includes all the formal definitions and the source code of the presented case study and other case studies can be found in [12].

2   Related Work

Several attempts have been made to check that a protocol role implementation is correct w.r.t. its specification which can be grouped in four main categories: (1) Model Driven Development (MDD); (2) Static Code Verification; (3) Refinement Types; (4) Online Monitoring and Intrusion Detection Systems (IDSs).

The first approach consists of designing and verifying a formal, high-level model of a security protocol and to semi-automatically derive an implementation that satisfies the same security properties of the formal model [8,9,13]. However, it has the drawback of not handling legacy implementations of security protocols.

The second approach starts from the source code of an existing implementation, and extracts a formal model which is verified for the desired security properties [5,11]. In principle, this approach can deal with legacy implementations, but their source code must be available, which is not always the case.

The third approach proves security properties of an implementation by means of a special kind of type checking on its source code [4]. Working on the source code, it shares the same advantages and drawbacks of the second approach.

The fourth approach comes in two versions. With online monitoring, the source code of an existing implementation is instrumented with assertions: program execution is stopped if any assertion fails at runtime [3]. Besides requiring the source code, the legacy implementation must be substituted with the instrumented one, which may not always be the case. IDSs are systems that monitor network traffic and compare it against known attack patterns, or expected average network traffic. By working on averages, in order not to miss real attacks, IDSs often report false positive warnings. In order to reduce them, sometimes the source code of the monitored implementation is also instrumented [10], sharing the same advantages and drawback of online monitoring.

Another branch of research focused on security wrappers for legacy implementations. In [1], a formal approach that uses security wrappers as firewalls with an encrypting tunnel is described. Any communication that crosses some security boundary is automatically and transparently encrypted by the wrappers. That is, the wrappers *add* security to a distributed legacy system. In our approach, the monitor *enforces* the security already present in the system. Technically, our approach derives a monitor based on the security requirements already present in the legacy system, instead of adding a boilerplate layer of security.

Analogously, in [7] wrappers are used, among other things, to transparently encrypt local files accessed by library calls. However, distributed environments are not taken into account. Finally, in [17] wrappers are used to harden software libraries. However, cryptography and distributed systems are not considered, and the approach is test-driven, rather than formally based.

3   Formal Background

3.1   Network Model

Many network models have been proposed in the Dolev-Yao setting. For example, sometimes the network is represented as a separate process [16]; the attacker is connected to this network, and can eavesdrop, drop and modify messages, or forge new ones. In other cases, the attacker is the medium [15], and honest agents can only communicate *through* the attacker. Even more detailed network models have been developed [19], where some nodes may have direct, private secured communication with other nodes, while still also being able to communicate through insecure channels, controlled by the attacker.

In general, it is not trivial to show that all of these models are equivalent in a Dolev-Yao setting, furthermore different network models and agents granularity justify different positions of the monitor with respect to the monitored agent, affecting the way the monitor is actually implemented. In this paper, we focus on a simple scenario that is usually found in practice, and is depicted in figure 2(a): the attacker is the medium, and every protocol agent communicates over a single insecure channel c, and private channels are not allowed. Moreover, agents are sequential and non-recursive.

Let us define A as the (correct) model of the agent to be monitored, and M_A as the model of its monitor. When the monitor is present, A communicates

(a) Agents A and B with the attacker.   **(b)** Agent A monitored by M_A and the attacker.

Fig. 2. The network model

with M_A only, through the use of a private channel c_{AM}, while M_A is directly connected to the attacker by channel c, as depicted in figure 2(b). The dashed box denotes that A and M_A run in the same environment, for example they run on the same system with same privileges. Note that in A channel c is in fact renamed to c_{AM}.

3.2 The Spi Calculus

In this paper, the formal models are expressed in Spi Calculus [2]. Spi Calculus is amenable for our approach because it is a domain specific language tailored at expressing the behaviour of single security protocol agents, where checks on received data must be explicitly specified. Thus, from the Spi Calculus specifications of protocol agents, the $a2m$ function can derive precise and complete specifications of their monitors.

Briefly, a Spi Calculus specification is a system of concurrent processes that operate on untyped data, called terms. Terms can be exchanged between processes by means of input/output operations. Table 1(a) contains the terms defined by the Spi Calculus, while table 1(b) shows the processes.

A name n is an atomic value, and a pair (M, N) is a compound term, composed of the terms M and N. The 0 and $suc(M)$ terms represent the value of zero and the logical successor of some term M, respectively. A variable x represents any

Table 1. Spi Calculus grammar

(a) Spi Calculus terms.

$L, M, N ::=$	terms
n	name
(M, N)	pair
0	zero
$suc(M)$	successor
x	variable
$M^{\sim}$	shared-key
$\{M\}_N$	shared-key encryption
$H(M)$	hashing
M^+	public part
M^-	private part
$\{[M]\}_N$	public-key encryption
$[\{M\}]_N$	private-key signature

(b) Spi Calculus processes.

$P, Q, R ::=$	processes
$\overline{M}\langle N\rangle.P$	output
$M(x).P$	input
$P \mid Q$	composition
$!P$	replication
$(\nu n)\,P$	restriction
$[M \text{ is } N]\,P$	match
0	nil
$let\ (x, y) = M\ in\ P$	pair splitting
$case\ M\ of\ 0 : P\ suc(x) : Q$	integer case
$case\ L\ of\ \{x\}_N\ in\ P$	shared-key decryption
$case\ L\ of\ \{[x]\}_N\ in\ P$	decryption
$case\ L\ of\ [\{x\}]_N\ in\ P$	signature check

term, and it can be bound once to the value of another term. If a variable or a name is not bound, then it is free. The $M^\sim$ term represents a symmetric key built from key material M, and $\{M\}_N$ represents the encryption of the plaintext M with the symmetric key N, while $H(M)$ represents the result of hashing M. The M^+ and M^- terms represent the public and private part of the keypair M respectively, while $\{[M]\}_N$ and $[\{M\}]_N$ represent public key and private key asymmetric encryptions respectively.

Informally, the $\overline{M}\langle N\rangle.P$ process sends message N on channel M, and then behaves like P, while the $M(x).P$ process receives a message from channel M, and then behaves like P, with x bound to the received term in P. A process P can perform an input or output operation iff there is a reacting process Q that is ready to perform the dual output or input operation. Note, however, that processes run within an environment (the Dolev-Yao attacker) that is always ready to perform input or output operations. Composition $P|Q$ means parallel execution of processes P and Q, while replication $!P$ means an unbounded number of instances of P run in parallel. The restriction process $(\nu n)P$ indicates that n is a fresh name (i.e. not previously used, and unknown to the attacker) in P. The match process executes like P, if M equals N, otherwise is stuck. The nil process does nothing. The pair splitting process binds the variables x and y to the components of the pair M, otherwise, if M is not a pair, the process is stuck. The integer case process executes like P if M is 0, else it executes like Q if M is $suc(N)$ and x is bound to N, otherwise the process is stuck. If L is $\{M\}_N$, then the shared-key decryption process executes like P, with x bound to M, else it is stuck, and analogous reasoning holds for the decryption and signature check processes.

The assumption that A is a sequential process, means that composition and replication are never used in its specification.

4   The Monitor Generation Function

The $a2m$ function translates a sequential protocol role specification into a monitor specification for that role; formally, $M_A \triangleq a2m(A)$. For brevity, $a2m$ is only informally presented here, by means of a running example. Formal definitions can be found in [12].

Before introducing the function, the concepts of *known* and *reconstructed* terms are given. For any Spi Calculus state, a term T is said to be *known* by the monitor through variable $\_T$, iff $\_T$ is bound to T. This can happen either because the implementation of M_A has access to the agent's memory location where T is stored; or because T can be read from a communication channel, and M_A stores T in variable $\_T$. A compound term T (that is not a name or a variable) is said to be *reconstructed*, if all its subterms are known or reconstructed. For example, suppose M is known through $\_M$ and $H(N)$ is known through $\_H(N)$. It is the case that $(H(N), M)$ is reconstructed by $(\_H(N), \_M)$. Note that, as terms become known, other terms may become reconstructed too. In the example given above, if M was not known, then it was not possible to reconstruct $(H(N), M)$;

```
1a: A(M,k) :=                    1m: MA(k,_H(M)) :=
2a:   cAM<{M}k>.                 2m:   cAM(_{M}k).
                                 3m:   case _{M}k of {_M}k in
                                 4m:   [_H(M) is H(_M)]
                                 5m:   c<_{M}k>.
3a:   cAM(x).                    6m:   c(x).
4a:   [x is H(M)]                7m:   [x is _H(M)]
5a:   0                          8m:   cAM<x>.
                                 9m:   0
```

(a) Agent A specification.   (b) Monitor specification derived from agent A one.

Fig. 3. Example specification of agent A along with its derived monitor M_A

however, if later M became known (for example, because it was sent over a channel), then $(H(N), M)$ would become reconstructed.

Note that the monitor implementation presented in this paper does not enforce that nonces are actually different for each protocol run. To enable this, the monitor should track all previously used values, in order to ensure that no value is used twice. Especially in the online mode, this overhead may not be acceptable. In order to drop this check, it is needed to assume that the random value generator in the monitored agent is correctly implemented. Also note that there may be cases where the monitor has not enough information to properly check protocol execution. These cases are recognised by the $a2m$ function, so that an error state is reached, and no unsound monitor is generated.

The $a2m$ function behaviour is now described by means of a running example. Agent A sends some data M encrypted by the key k to the other party, and expects to receive the hash of the plaintext, that is $H(M)$. Note that the example focuses on the way the $a2m$ function operates, rather than on monitoring a security protocol, so no security claim is meant to be made on this protocol. Figure 3(a) shows the specification for agent A, and figure 3(b) its derived monitor specification M_A. Here, an ASCII syntax of Spi Calculus is used: the 'ν' symbol is replaced by the '@' symbol, and the overline in the output process is omitted (input and output processes can still be distinguished by the different brackets).

At line 1a the agent A process is declared: it has two free variables, a message M and a symmetric key k. At line 2a A sends the encryption of M with key k. Then, at line 3a it receives a message that is stored into variable x, and, at line 4a, the received message is checked to be equal to the hashing of M: if this is the case, the process correctly terminates.

At line 1m, the monitor M_A is declared: to make this example significant, it is assumed that in the initial state the key k used by A is known by the monitor (through the variable k), while M is not known (for example, because the monitor cannot access those data); however $H(M)$ is known through $\_H(M)$, that is the monitor has access to the memory location where $H(M)$ is stored, and this value is bound to the variable $\_H(M)$ in the monitor.

When line 2a is translated by $a2m$, lines 2m–5m are produced. The data sent by A are received by the monitor at line 2m, and stored in variable $\_\{M\}_k$. Afterwards, some checks on the received value are added by the $a2m$ function. In general, each time a new message is received from the monitored application, it or its parts are checked against their expected (known or reconstructed) values. In this case, since $\{M\}_k$ is not known (by hypothesis) or reconstructed (because M is not known or reconstructed), it cannot be directly compared against the known or reconstructed value, so it is exploded into its components. As $\{M\}_k$ is an encryption and k is known, the decryption case process is generated at line 3m, binding $\_M$ to the value of the plaintext, that should be M. Since M is not known or reconstructed, and it is a name, $\_M$ cannot be dissected any more; instead, M becomes known through $\_M$, in other words, the term stored in $\_M$ is assumed by the monitor to be the correct term for M. Note that, before M was known through $\_M$, $H(M)$ was known through $\_H(M)$, but it was not reconstructed. After the assignment of $\_M$, $H(M)$ becomes reconstructed by $H(\_M)$ too. The match process at line 4m ensures that known and reconstructed values for the same term are actually the same.

After all the possible checks are performed on the received data, they are forwarded to the attacker at line 5m. Then, line 3a is translated into line 6m. When translating an input process, the monitor receives message x from the attacker on behalf of the agent and buffers it; x is said to be known through x itself. Then the monitor behaves according to what is done by the agent (usually checks on the received data, as it is the case in the running example). The received message stored in x is not forwarded to A immediately, because this could lead A to receive some malicious data, that could for example enable some denial of service attack. Instead, the received data are buffered, and will be forwarded to A only when necessary: that is when the process should end (**0** case), or when some output data from A are expected.

Line 4a is then translated into line 7m, and finally, line 5a is translated into lines 8m and 9m. First, all buffered data (x in this case) are forwarded to A, then the monitor correctly ends.

5   An SSL Server Monitor Example

5.1   Monitor Specification

As shown in figure 1, in order to get the monitor specification, a Spi Calculus specification of the server role for the SSL protocol is needed. The full SSL protocol is rather complex: many scenarios are possible, and different sets of cryptographic algorithms (called *ciphersuites*) can be negotiated. For simplicity, this example considers only one scenario of the SSL protocol version 3.0, where the same cipher suite is always negotiated. Despite these simplifications, we believe that the considered SSL fragment is still significant, and that adding full SSL support would increase the example complexity more than its significance.

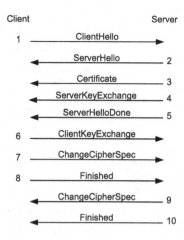

Fig. 4. Typical SSL scenario

In this paper, the chosen scenario requires the server to use a DSA certificate for authentication and data signature. Although RSA certificates are more common in SSL, using a DSA certificate allowed us to stress a bug in the OpenSSL implementation, showing that the monitor can actively drop malicious sessions that would be otherwise accepted as genuine by the flawed OpenSSL implementation. The more common RSA scenario has been validated through a dedicated case-study. However, it is not reported here for brevity; it can be found in [12].

The SSL scenario considered in this example is depicted informally in figure 4, while figure 5 shows a possible Spi Calculus specification of a server for the chosen scenario. The ASCII syntax of Spi Calculus is used in figure 5. Also, in order to model the Diffie-Hellman (DH) key exchange, the $EXP(L, M, N)$ term is added, which expresses the modular exponentiation L^M mod N, along with the equation $EXP(EXP(g, a, p), b, p) = EXP(EXP(g, b, p), a, p)$.

To make the specification more readable, lists of terms like (A, B, C) are added as syntactic sugar, and they are translated into left associated nested pairs, like $((A, B), C)$; a `rename n = M in P` process is introduced too, that renames the term M to n and then behaves like P.

The first message is the ClientHello, sent from the client to the server. It contains the highest protocol version supported by the client, a random value, a session ID for session resuming, and the set of supported cipher suites and compression methods. In the server specification, the ClientHello message is received and split into its parts at line 2S. In the chosen scenario, the client should send at least 3.0 as the highest supported protocol version, and it should send 0 as session ID, so that no session resuming will be performed. Moreover, the client should at least support the always-negotiated cipher suite, namely SSL_DHE_DSS_3DES_EDE_CBC_SHA, with no compression enabled. All these constraints are checked at lines 3S–4S.

```
1S  Server() :=
2S    c(c_hello).  let (c_version,c_rand,c_SID,c_ciph_suite,c_comp_method) = c_hello in
3S    [ c_version is THREE_DOT_ZERO ]  [ c_SID is ZERO ]
4S    [ c_ciph_suite is SSL_DHE_DSS_3DES_EDE_CBC_SHA ]  [ c_comp_method is comp_NULL ]
5S    (@s_rand)  (@SID)
6S    rename S_HELLO = (THREE_DOT_ZERO,s_rand,SID,SSL_DHE_DSS_3DES_EDE_CBC_SHA,comp_NULL) in
7S    (@DH_s_pri)  rename DH_s_pub = EXP(DH_Q,DH_s_pri,DH_P) in
8S    rename S_KEX = ((DH_P,DH_Q,DH_s_pub),[{H(c_rand,s_rand,(DH_P,DH_Q,DH_s_pub))}]s_PriKey) in
9S    c<S_HELLO,S_CERT,S_KEX,S_HELLO_DONE>.
10S   c(c_kex).  let (c_kexHead,DH_c_pub) = c_kex in  rename PMS = EXP(DH_c_pub,DH_s_pri,DH_P) in
11S   rename MS = H(PMS,c_rand,s_rand) in  rename KM = H(MS,c_rand,s_rand) in
12S   rename c_w_IV = H(KM,C_WRITE_IV) in  rename s_w_IV = H(KM,S_WRITE_IV) in
13S   c(c_ChgCipherSpec).  [ c_ChgCipherSpec is CHG_CIPH_SPEC ]
14S   c(c_encrypted_Finish).  case c_encrypted_Finish of {c_Finish_and_MAC}(KM,C_WRITE_KEY)~ in
15S   let (c_Finish,c_MAC) = c_Finish_and_MAC in  [ c_MAC is H((KM,C_MAC_SEC)~,c_Finish) ]
16S   let (final_Hash_MD5, final_Hash_SHA) = c_Finish in
17S   [ final_Hash_MD5 is H((c_hello,S_HELLO,S_CERT,S_KEX,S_HELLO_DONE,c_kex),C_ROLE,MS,MD5) ]
18S   [ final_Hash_SHA is H((c_hello,S_HELLO,S_CERT,S_KEX,S_HELLO_DONE,c_kex),C_ROLE,MS,SHA) ]
19S   c<CHG_CIPH_SPEC>.
20S   rename DATA = (c_hello,S_HELLO,S_CERT,S_KEX,S_HELLO_DONE,c_kex,c_Finish) in
21S   rename S_FINISH = (H(DATA,S_ROLE,MS,MD5),H(DATA,S_ROLE,MS,SHA)) in
22S   (@pad)  c<{S_FINISH,H((KM,S_MAC_SEC)~,S_FINISH),pad}(KM,S_WRITE_KEY)~>.
23S   0
```

Fig. 5. A possible Spi Calculus specification of an SSL server

In the second message the server replies by sending its ServerHello message, that contains the chosen protocol version, a random value, a fresh session ID and the chosen cipher suite and compression method. The server random value and the fresh session ID are generated at line 5S, then the ServerHello message is declared at line 6S. Again, in the chosen scenario, the server chooses protocol version 3.0, and always selects the SSL_DHE_DSS_3DES_EDE_CBC_SHA cipher suite, with no compression enabled. Then the server sends the Certificate message to the client: in the chosen scenario, this message contains a DSA certificate chain for the server's public key, that authenticates the server.

In the fourth message, named ServerKeyExchange, the server sends the DH key exchange parameters to the client, and digitally signs them with its public key. In the server specification, the DH server secret value DH_s_pri and the corresponding public value DH_s_pub are computed at line 7S. Then, at line 8S, the ServerKeyExchange message is declared: it consists of the server DH parameters, along with a digital signature of the DH parameters and the client and server random values, in order to ensure signature freshness.

The fifth message is the ServerHelloDone. It contains no data, but signals the client that the server ended its negotiation part, so the client can move to the next protocol stage. In the server specification, these four messages are sent all at once at line 9S.

In the sixth message, the client replies with the ClientKeyExchange message, that contains the client's DH public value. Note that there is no digital signature in this message, since the client is not authenticated. In the server specification the ClientKeyExchange is received at line 10S, where the payload is split from the message header too. Both client and server derive a shared secret from the DH key exchange. This shared secret is called Premaster Secret (PMS), and it is used by both parties to derive some shared secrets used for symmetric encryption of application data. The PMS is computed by the server at line 10S. By applying an SSL custom hashing function to the PMS and the client and server random data, both client and server can compute the same Master Secret (MS). The bytes of the MS are then extended (again by using a custom SSL hashing algorithm) to as many byte as required by the negotiated ciphersuite, obtaining the Key Material (KM) (line 11S). Finally, different subsets of bytes of the KM are used as shared secrets and as initialization vectors (IVs). Note that IVs, that are extracted at line 12S, are never referenced in the specification. They will be used as cryptographic parameters for subsequent encryptions, during the code generation step, explained in section 5.2.

The seventh message is the ChangeCipherSpec, received and checked at line 13S. This message contains no data, but signals the server that the client will start using the negotiated cipher suite from the next message on.

The client then sends its Finished message. Message Authentication Code (MAC) and encryption are applied to the Finished message sent by the client, as the client already sent its ChangeCipherSpec message. The client Finished message is received and decrypted at line 14S. The decryption key used (KM,C_WRITE_KEY)~ is obtained by creating a shared key, starting from the key material KM and a marker C_WRITE_KEY that indicates which portion of the KM to use. At line 15S the MAC is extracted from the plaintext, and verified. The unencrypted content of the Finished message contains the final hash, that hashes all relevant session information: all exchanged messages (excluding the Change-CipherSpec ones) and the MS are included in the final hash, plus some constant data identifying the protocol role (client or server) that sent the final hash. In fact, the Finished message includes two versions of the same final hash, one using the MD5 algorithm, and one using the SHA-1 algorithm. Both versions of the final hash are extracted and checked at lines 16S–18S. As Spi Calculus does not support different algorithms for the same hash, they are distinguished by a marker (MD5 and SHA respectively) as the last hash argument, making them syntactically different.

Then the server sends its ChangeCipherSpec message to client (line 19S), and its Finished message, that comes with MAC and encryption too (lines 20S–22S). Encryption requires random padding to align the plaintext length to the cipher block size. This random padding must be explicitly represented in the server specification, so that the monitor can recognise and discard it, and only check the real plaintext. Otherwise the monitor would try to locally reconstruct the encryption, but it would always fail, because it could not guess the padding. The protocol handshake is now complete, and next messages will contain secured application data.

In order to verify any security property on this specification, the full SSL specification, including the client and protocol sessions instantiations is required. However, this is outside the scope of this paper; SSL security properties have already been verified, for example, by the AVISPA project [18]. Here it is assumed instead that the specification of the server is correct, and thus secure, so that the monitoring approach can be shown.

The $a2m$ function described in section 4 is applied on the server specification, in order to obtain the online monitor specification for the server role. For brevity, the resulting specification is not shown here. It can be found, along with more implementation details, in [12]. It is assumed that the monitor has access to the server private DH value, which is then *known*, while it is not able to read the freshly generated server random value s_rand, the session ID SID and the random padding which are then not known nor reconstructed at generation time. Often, the server will generate a fresh DH private value for each session, and it will usually only store it in memory. In general, with some effort the monitor will be able to directly read this secret from the legacy application memory, without the need of the source code. Nevertheless, in a testing environment, if the source code of the monitored application happens to be available, it is possible to patch the monitored application, so that it explicitly communicates the DH private value to the monitor. Indeed, this is reasonable because the monitor is part of the trusted system, and is actually more trusted than the monitored application.

5.2   Monitor Implementation

The source code of the monitor implementation can be found in [12]. In order to generate the monitor implementation, the spi2java MDD framework is used [13]. Briefly, spi2java is a toolchain that, starting from the formal Spi Calculus specification of a security protocol, semi-automatically derives an interoperable Java implementation of the protocol actors. In the first place, spi2java was designed to generate security protocol actors, rather than monitors. In this paper, we originally reuse spi2java to generate a monitor.

In order to generate an executable Java implementation of a Spi Calculus specification, some details that are not contained in the Spi Calculus specification must be added. That is, the Spi Calculus specification must be refined, before it can be translated into a Java application.

As shown in figure 1, the spi2java framework assists the developer during the refinement and code generation steps. The spi2java refiner is used to automatically infer some refinement information from the given specification. All inferred information is stored into an eSpi (extended Spi Calculus) document, which is coupled with the Spi Calculus specification. The developer can manually refine the generated eSpi document; the manually refined eSpi document is passed back to the spi2java refiner, that checks its coherence against the original Spi Calculus specification, and possibly infers new information from the user given one. This iterative refinement step can be repeated until the developer is satisfied with the obtained eSpi document, but usually one iteration is enough.

The obtained eSpi document and the original Spi Calculus specification are passed to the spi2java code generator that automatically outputs the Java code implementing the given specification. The generated code implements the "protocol logic", that is the code that simulates the Spi Calculus specification by coordinating input/output operations, cryptographic primitives and checks on received data. Dealing with Java sockets or the Java Cryptographic Architecture (JCA) is delegated to the SpiWrapper library, which is part of the spi2java framework. The SpiWrapper library allows the generated code to be compact and readable, so that it can be easily mapped back to the Spi Calculus specification. For example, the monitor specification corresponding to line 2S of the server specification in figure 5 is translated as

```
/* c_0(c_hello_1). */
Pair c_hello_1 = (Pair) c_0.receive(new PairRecvClHello());
```

(each Spi Calculus term name is mangled to make sure there is a unique Java identifier for that term). To improve readability, the spi2java code generator outputs the translated Spi Calculus process as a Java comment too. In this example, the Java variable c_0 has type `TcpIpChannel`, which is a Java class included in the SpiWrapper library implementing a Spi Calculus channel using TCP/IP as transport layer. This class offers the `receive` method that allows the Spi Calculus input process to be easily implemented, by internally dealing with the Java sockets. The `c_hello_1` Java variable has type `Pair`, which implements the Spi Calculus pair. The `Pair` class offers the `getLeft` and `getRight` methods, allowing a straightforward implementation of the pair splitting process. The spi2java translation function is proven sound in [14].

In order to get interoperable implementations, the SpiWrapper library classes only deal with the *internal* representation of data. By extending the SpiWrapper classes, the developer can provide custom marshalling functions that transform the internal representation of data into the external one.

In the SSL monitor case study, a two-tier marshalling layer has been implemented. Tier 1 handles the Record Layer protocol of SSL, while tier 2 handles the upper layer protocols. When receiving a message from another agent, tier 1 parses one Record Layer message from the input stream, and its contained upper layer protocol messages are made available to tier 2. The latter implements the real marshalling functions, for example converting US-ASCII strings to and from Java String objects. Analogous reasoning applies when sending a message. The marshalling layer functions only check that the packet format is correct. No control on the payload is needed: it will be checked by the automatically generated protocol logic.

The SSL protocol defines custom hashing algorithms, for instance to compute the MS from the PMS, or to compute the MAC value. For each of them, a corresponding SpiWrapper class has been created, implementing the custom algorithm. Moreover, the spi2java framework has been extended to support the modular exponentiation, so that DH key exchange can be supported.

Finally, it is worth pointing out some details about the IVs used by cryptographic operations (declared in the server specification at line 12S). For each term of the form $\{M\}_K$, the eSpi document allows its cryptographic algorithm (such as DES, 3DES, AES) and its IV to be specified. However, the IV is only known at run time. The spi2java framework allows cryptographic algorithms and parameters to be resolved either at compile time or at run time. If the parameter is to be resolved at compile time, the value of the parameter must be provided (e.g. AES for the symmetric encryption algorithm, or a constant value for the IV). If the parameter is to be resolved at run time, the identifier of another term of the Spi Calculus specification must be provided: the parameter value will be obtained by the content of the referred term, during execution. In the SSL case study, this feature is used for the IVs. For example, the {c_Finish_and_MAC}(KM,C_WRITE_KEY)~ term uses the H(KM,C_WRITE_IV) term as IV. Technically, this feature enables support for cipher suite negotiation. However, as stated above, this would increase the specification complexity more than it would increase its significance, and is left for future work.

5.3   Experimental Results

The monitor has been coupled in turn with three different SSL server implementations, namely OpenSSL[1] version 0.9.8j, GnuTLS[2] version 2.4.2 and JESSIE[3] version 1.0.1.

Since the online monitoring paradigm is used in this case study, the monitor is accepting connections on the standard SSL port (443), while the real server is started on another port (4433). Each time a client connects to the monitor, the latter opens a connection to the real server, starting data checking and forwarding, as explained above.

It is worth noting that switching the server implementation is straightforward. In the testing scenario, assuming that the server communicates its private DH value to the monitor, it is enough to shut down the running server implementation, and to start the other one; the monitor implementation remains the same, and no action on the monitor is required. Otherwise, it is enough to restart the monitor too, enabling the correct plugin that gathers the private DH value from the legacy application memory. In other words, in a production scenario, the same monitor implementation can handle several different legacy server implementations; in the monitor, the only server-dependent part is the plugin that reads the DH secret value from the server application memory.

In order to generate protocol sessions, three SSL clients have been used with each server; namely the OpenSSL, GnuTLS, and JESSIE clients. During experiments, the monitor helped in spotting a bug in the JESSIE client: This client always sends packet of the SSL 3.1 version (better known as TLS 1.0), regardless of the negotiated version, that is SSL 3.0 in our scenario. The monitor correctly rejected all JESSIE client sessions, reporting the wrong protocol version.

When the OpenSSL or GnuTLS clients are used, the monitor correctly operates with all the three servers. In particular, safe sessions are successfully handled; conversely, when exchanged data are manually corrupted, they are recognized by the monitor and the session is aborted: corrupted data are never forwarded to the intended recipient.

In order to estimate the impact on performances of the online monitoring approach, execution times of correctly ended protocol sessions with and without the monitor have been measured. Thus, performances regarding the JESSIE

---

[1] Available at: http://www.openssl.org/

[2] Available at: http://www.gnu.org/software/gnutls/

[3] Available at: http://www.nongnu.org/jessie/

Table 2. Average execution times for protocol runs with and without monitoring

Client	Server	No Monitor [s]	Monitor [s]	Overhead [s]	Overhead [%]
OpenSSL	OpenSSL	0.032	0.113	0.081	253.125
GnuTLS	OpenSSL	0.108	0.132	0.024	22.253
OpenSSL	GnuTLS	0.073	0.128	0.056	76.552
GnuTLS	GnuTLS	0.109	0.120	0.011	10.313
OpenSSL	JESSIE	0.158	0.172	0.014	8.986
GnuTLS	JESSIE	0.144	0.148	0.004	2.788

client are not reported, as no correct session could be completed, due to the discovered bug. That is, the measured performances all correspond to valid executions of the protocol only. Communication between client, server and monitor happened over local sockets, so that no random network delays could be introduced; moreover system load was constant during test execution. Table 2 shows the average execution times for different client-server pairs, with and without monitor enabled. For each client-server pair, the average execution times have been computed over ten protocol runs. Columns "No Monitor" and "Monitor" report the average execution times, in seconds, without and with monitoring enabled respectively. When monitoring is not enabled, the clients directly connect to the server on port 4433. The "Overhead" columns show the overhead introduced by the monitor, in seconds and in percentage respectively. In four cases out of six, the monitor overhead is under 25 milliseconds. From a practical point of view, a client communicating through a real distributed network could hardly tell whether a monitor is present or not, since network times are orders of magnitude higher. On the other hand, in the worst cases online monitoring can slow down the server machine up to 2.5 times. Whether this overhead is acceptable on the server side depends on the number of sessions per seconds that must be handled. If the overhead is not acceptable, the offline monitoring paradigm can still be used.

The OpenSSL security flaw. Recently, the client side of the OpenSSL library prior to version 0.9.8j has been discovered flawed, such that in principle it could treat a malformed DSA certificate as a good one rather than as an error.[4] By inspecting the flawed code, we were able to forge such malformed certificate that exploited the affected versions. This malformed DSA certificate must have the q parameter one byte longer than expected. Up to our knowledge, this is the first documented and repeatable exploit for this flaw.

Without monitoring enabled, we generated protocol sessions between an SSL server sending the offending certificate, and both OpenSSL clients version 0.9.8i (flawed) and 0.9.8j (fixed). By using the `-state` command line argument, it is possible to conclude that the 0.9.8i version completes the handshake by reaching the "read finished A" state (after message 10 in figure 4); while the 0.9.8j version correctly reports an "handshake failure" error at state "read server certificate A", that is immediately after message 3 in figure 4.

---

[4] http://www.openssl.org/news/secadv_20090107.txt

When monitoring is enabled, the malformed server certificate is passed to the monitor as an input parameter, that is, the server certificate is *known* by the monitor. In this case the monitor actually refuses to start. Indeed, when loading the server certificate, the monitor spots that it is malformed, and does not allow any session to be even started. If we drop the assumption that the monitor knows the server certificate, then the monitor starts, and checks the server certificate when it is received over the network. During these checks, the malformed certificate is found, and the session is dropped, before the server Certificate message is forwarded to the client. This prevents the aforecited flaw to be exploited on OpenSSL version 0.9.8i.

6   Conclusion

The paper shows a formally-based yet practical methodology to design, develop and deploy monitors for legacy implementations of security protocols, without the need to modify the legacy implementations or to analyse their source code. To our knowledge, this is the first work that allows legacy implementations of security protocol agents to be black-box monitored.

This paper introduces a function that, given the specification of a security protocol actor, automatically generates the specification of a monitor that stops incorrect sessions, without rising false positive alarms. From the obtained monitor specification, an MDD approach is used to generate a monitor implementation; for this purpose, the spi2java framework has been originally reused, and some of its parts enhanced.

Finally, the proposed methodology has been validated by implementing a monitor starting from the server role of the widely used SSL protocol. Core insights gained from conducting the SSL case study include that the same generated monitor implementation can in fact monitor several different SSL server implementations against different clients, in a black-box way. The only needed information is the private Diffie-Hellman key used by the server, in order to check message contents. Moreover, by reporting session errors, the monitor effectively helped us in finding a bug in an open source SSL client implementation.

The "online" monitoring paradigm proved useful in avoiding protocol violations, for example by stopping malicious data that would have otherwise exploited a known flaw of the widely deployed OpenSSL client. The overhead introduced by the monitor to check and forward messages is usually negligible. If the overhead is not acceptable, this paper also proposes an "offline" monitoring strategy that has no overhead and can still be useful to timely discover protocol attacks.

As future work, a general result about soundness of the monitor specification generating function would be useful. The soundness property should show that the generated monitor specification actually forwards only (and all) the protocol sessions that would be accepted by the agent's verified specification. Together with the soundness proofs of the spi2java framework, this would produce a sound monitor implementation, directly from the monitored agent's specification.

References

1. Abadi, M., Fournet, C., Gonthier, G.: Secure implementation of channel abstractions. Information and Computation 174(1), 37–83 (2002)
2. Abadi, M., Gordon, A.D.: A calculus for cryptographic protocols: The Spi Calculus. Digital Research Report 149 (1998)
3. Bauer, A., Jürjens, J.: Security protocols, properties, and their monitoring. In: International Workshop on Software Engineering for Secure Systems, pp. 33–40 (2008)
4. Bengtson, J., Bhargavan, K., Fournet, C., Gordon, A.D., Maffeis, S.: Refinement types for secure implementations. In: Computer Security Foundations Symposium, pp. 17–32. IEEE, Los Alamitos (2008)
5. Bhargavan, K., Fournet, C., Gordon, A.D., Tse, S.: Verified interoperable implementations of security protocols. In: Computer Security Foundations Workshop, pp. 139–152 (2006)
6. Dolev, D., Yao, A.C.C.: On the security of public key protocols. IEEE Transactions on Information Theory 29(2), 198–207 (1983)
7. Fraser, T., Badger, L., Feldman, M.: Hardening COTS software with generic software wrappers. In: IEEE Symposium on Security and Privacy, pp. 2–16 (1999)
8. Hubbers, E., Oostdijk, M., Poll, E.: Implementing a formally verifiable security protocol in Java Card. In: Hutter, D., Müller, G., Stephan, W., Ullmann, M. (eds.) Security in Pervasive Computing. LNCS, vol. 2802, pp. 213–226. Springer, Heidelberg (2004)
9. Jeon, C.W., Kim, I.G., Choi, J.Y.: Automatic generation of the C# code for security protocols verified with Casper/FDR. In: International Conference on Advanced Information Networking and Applications, pp. 507–510 (2005)
10. Joglekar, S.P., Tate, S.R.: ProtoMon: Embedded monitors for cryptographic protocol intrusion detection and prevention. Journal of Universal Computer Science 11(1), 83–103 (2005)
11. Jürjens, J., Yampolskiy, M.: Code security analysis with assertions. In: IEEE/ACM International Conference on Automated Software Engineering, pp. 392–395 (2005)
12. Pironti, A., Jürjens, J.: Online resources about black-box monitoring, http://alfredo.pironti.eu/research/projects/monitoring/
13. Pironti, A., Sisto, R.: An experiment in interoperable cryptographic protocol implementation using automatic code generation. In: IEEE Symposium on Computers and Communications, pp. 839–844 (2007)
14. Pironti, A., Sisto, R.: Provably correct Java implementations of Spi Calculus security protocols specifications. Computers & Security (2009) (in press)
15. Roscoe, A.W., Hoare, C.A.R., Bird, R.: The Theory and Practice of Concurrency. Prentice Hall PTR, Englewood Cliffs (1997)
16. Schneider, S.: Security properties and CSP. In: IEEE Symposium on Security and Privacy, pp. 174–187 (1996)
17. Süßkraut, M., Fetzer, C.: Robustness and security hardening of COTS software libraries. In: IEEE/IFIP International Conference on Dependable Systems and Networks, pp. 61–71 (2007)
18. Viganò, L.: Automated security protocol analysis with the AVISPA tool. Electronic Notes on Theoretical Computer Science 155, 61–86 (2006)
19. Voydock, V.L., Kent, S.T.: Security mechanisms in high-level network protocols. ACM Computing Surveys 15(2), 135–171 (1983)

Secure Code Generation for Web Applications

Martin Johns[1,2], Christian Beyerlein[3], Rosemaria Giesecke[1],
and Joachim Posegga[2]

[1] SAP Research – CEC Karlsruhe
{martin.johns,rosemaria.giesecke}@sap.com
[2] University of Passau, Faculty for Informatics and Mathematics, ISL
{mj,jp}@sec.uni-passau.de
[3] University of Hamburg, Department of Informatics, SVS
9beyerle@informatik.uni-hamburg.de

Abstract. A large percentage of recent security problems, such as Cross-site Scripting or SQL injection, is caused by string-based code injection vulnerabilities. These vulnerabilities exist because of implicit code creation through string serialization. Based on an analysis of the vulnerability class' underlying mechanisms, we propose a general approach to outfit modern programming languages with mandatory means for explicit and secure code generation which provide strict separation between data and code. Using an exemplified implementation for the languages Java and HTML/JavaScript respectively, we show how our approach can be realized and enforced.

1 Introduction

The vast majority of today's security issues occur because of string-based code injection vulnerabilities, such as Cross-site Scripting (XSS) or SQL Injection. An analysis of the affected systems results in the following observation: All programs that are susceptible to such vulnerabilities share a common characteristics – They utilize the string type to assemble the computer code that is sent to application-external interpreters. In this paper, we analyse this vulnerability type's underlying mechanisms and propose a simple, yet effective extension for modern programming languages that reliably prevents the introduction of string-based code injection vulnerabilities.

1.1 The Root of String-Based Injection Vulnerabilities

Networked applications and especially web applications employ a varying amount of heterogeneous computer languages (see Listing 1), such as programming (e.g., Java, PHP, C#), query (e.g., SQL or XPATH), or mark-up languages (e.g., XML or HTML).

```
1  // embedded HTML syntax
2  out.write("<a href='http://www.foo.org'>go</a>");
3  // embedded SQL syntax
4  sql = "SELECT * FROM users";
```

Listing 1. Examples of embedded syntax

F. Massacci, D. Wallach, and N. Zannone (Eds.): ESSoS 2010, LNCS 5965, pp. 96–113, 2010.

This observation leads us to the following definition:

Definition 1 (Hosting/Embedded). *For the remainder of this paper we will use the following naming convention:*

- **Hosting language:** *The language that was used to program the actual application (e.g., Java).*
- **Embedded language:** *All other computer languages that are employed by the application (e.g., HTML, JavaScript, SQL, XML).*

Embedded language syntax is assembled at run-time by the hosting application and then passed on to external entities, such as databases or web browsers. String-based code injection vulnerabilities arise due to a mismatch in the programmer's intent while assembling embedded language syntax and the actual interpretation of this syntax by the external parser. For example, take a dynamically constructed SQL-statement (see Fig. 1.a). The application's programmer probably considered the constant part of the string-assembly to be the *code-*portion while the concatenated variable was supposed to add dynamically *data-*information to the query. The database's parser has no knowledge of the programmer's intent. It simply parses the provided string according to the embedded language's grammar (see Fig. 1.b). Thus, an attacker can exploit this discord in the respective views of the assembled syntax by providing data-information that is interpreted by the parser to consist partly of *code* (see Fig. 1.c).

In general all string values that are provided by an application's user on runtime should be treated purely as *data* and never be executed. But in most cases the hosting language does not provide a mechanism to explicitly generate embedded syntax. For this reason all embedded syntax is generated implicitly by string-concatenation and -serialization. Thus, the hosting language has no means to differentiate between user-provided dynamic *data* and programmer-provided embedded *code* (see Fig. 1.d). Therefore, it is the programmer's duty to make sure that all dynamically added *data* will not be parsed as *code* by the external interpreter. Consequently, if a flaw in the application's logic allows an inclusion of arbitrary values into a string segment that is passed as embedded syntax to an external entity, an attacker can succeed in injecting malicious *code*.

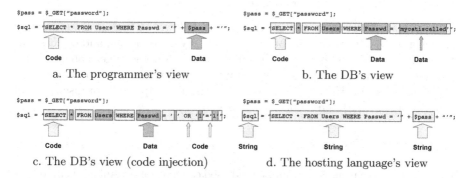

Fig. 1. Mismatching views on code

This bug pattern results in many vulnerability types, such as XSS, SQL injection, directory traversal, shell command injection, XPath injection or LDAP injection.

A considerable amount of work went into addressing this problem by approaches which track the flow of untrusted data through the application (see Sec. 4). However, with the exception of special domain solutions, the causing practice of using strings for code assembly has not been challenged yet.

1.2   The String Is the Wrong Tool!

As motivated above, assembly of computer code using the string datatype is fundamentally flawed. The string is a "dumb" container type for sequences of characters. This forces the programmer to create embedded syntax *implicitly*. Instead of clearly stating that he aims to create a `select`-statement which accesses a specific table, he has to create a stream of characters which hopefully will be interpreted by the database in the desired fashion. Besides being notoriously error-prone, this practice does not take full advantage of the programmer's knowledge. He knows that he wants to create a `select`-statement with very specific characteristics. But the current means do not enable him to state this intent *explicitly* in his source code.

1.3   Contributions and Paper Outline

In this paper we propose to add dedicated capabilities to the hosting language that

- allow secure creation of embedded syntax through
- enforcing explicit creation of code semantics by the developer and
- providing strict separation between embedded data and code.

This way, our method reliably prevents string-based code injection vulnerabilities. Furthermore, our approach is

- applicable for any given embedded language (e.g., HTML, JavaScript, SQL),
- can be implemented without significant changes in the hosting language,
- and mimics closely the current string-based practices in order to foster acceptance by the developer community.

The remainder of the paper is organised as follows. In Section 2 we describe our general objectives, identify our approach's key components, and discuss how these components can be realised. Then, in Section 3 we document our practical implementation and evaluation. For this purpose, we realised our method for the languages Java and HTML/JavaScript. We end the paper with a discussion of related work (Sec. 4) and a conclusion (Sec. 5).

2   Concept Overview

In this section, we propose our language-based concept for assembly of embedded syntax which provides robust security guarantees.

2.1   Keeping the Developers Happy

Before describing our approach, in this section we list objectives which we consider essential for any novel language-based methodology to obtain acceptance in the developer community.

Objectives concerning the hosting language: Foremost, the proposed concepts should not depend on the characteristics of a specific hosting language. They rather should be applicable for any programming language in the class of procedural and object-orientated languages[1]. Furthermore, the realisation of the concepts should also not profoundly change the hosting language. Only aspects of the hosting language that directly deal with the assembly of embedded code should be affected.

Objectives concerning the creation of embedded syntax: The specific design of every computer language is based on a set of paradigms that were chosen by the language's creators. These paradigms were selected because they fitted the creator's design goals in respect to the language's scope. This holds especially true for languages like SQL that were not designed to be a general purpose programming language but instead to solve one specific problem domain. Therefore, a mechanism for assembling such embedded syntax within a hosting language should aim to mimic the embedded language as closely as possible. If the language integration requires profound changes in the embedded syntax it is highly likely that some of the language's original design paradigms are violated. Furthermore, such changes would also cause considerable training effort even for developers that are familiar with the original embedded language.

Finally, string operations, such as string-concatenation, string-splitting, or search-and-replace, have been proven in practice to be a powerful tool for syntax assembly. Hence, the capabilities and flexibility of the string type should be closely mimicked by the newly introduced methods.

2.2   Key Components

We propose a reliable methodology to assemble embedded language syntax in which all code semantics are created explicitly and which provides a strict separation between data and code. In order to do so, we have to introduce three key components to the hosting language and its run-time environment: A dedicated datatype for embedded code assembly, a suitable method to integrate the embedded language's syntax into the hosting code, and an external interface which communicates the embedded code to the external entity (see Fig. 2).

Datatype: We have to introduce a novel datatype to the hosting language that is suitable to assemble/represent embedded syntax and that guarantees strict separation between data and code according to the developer's intent. In the context of this document we refer to such a datatype as **ELET**, which is short for *Embedded Language Encapsulation Type*.

---

[1] To which degree this objective is satisfiable for functional and logical programming languages has to be determined in the future.

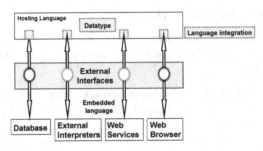

Fig. 2. Key components of the proposed approach

Language integration: To fill the ELET datatype, suitable means have to be
added to the hosting language which enable the developer to pass embedded syntax to the datatype. As motivated above these means should not
introduce significant changes to the embedded language's syntax.

External interface: For the time being, the respective external entities do not
yet expect the embedded code to be passed on in ELET form. For this reason,
the final communication step is still done in character based form. Therefore,
we have to provide an external interface which securely translates embedded
syntax which is encapsulated in an ELET into a character-based representation. Furthermore, all legacy, character-based interfaces to the external
entities have to be disabled to ensure our approach's security guarantees.
Otherwise, a careless programmer would still be able to introduce code injection vulnerabilities.

In the following sections we show how these three key components can be realised.

2.3   The Embedded Language Encapsulation Type (ELET)

This section discusses our design decisions concerning the introduced datatype
and explains our security claims.

As shown in Sec. 1.1, the root cause for string-based code injection vulnerabilities is that the developer-intended data/code-semantics of the embedded syntax
are not correctly enforced by the actual hosting code which is responsible for the
syntax assembly. Therefore, to construct a methodology which reliably captures
the developer's intent and provides strict separation between data and code, it
is necessary to clarify two aspects of embedded syntax assembly:

For one, an intended classification of syntax into *data* and *code* portions implies that an underlying segmentation of individual syntactical language elements
exist. Hence, the details of this segmentation have to be specified. Furthermore,
after this specification step, a rationale has to be built which governs the assignment of the identified language elements into the *data/code*-classes.

ELET-internal representation of the embedded syntax: The main design
goal of the ELET type is to support a syntax assembly method which bridges
the current gap between string-based code creation and the corresponding parsing process of the application-external parsers. For this reason, we examined the

mechanisms of parsing of source code: In general, the parsing is, at least, a two step process. First a lexical analysis of the input character stream generates a stream of tokens. Then, a syntactical analysis step matches these tokens according to the language's formal grammar and aligns them into tree structures such as parse-trees or abstract syntax-trees. The leafs of such trees are the previously identified tokens, now labeled according to their grammar-defined type (e.g., JavaScript defines the following token types: keyword-token, identifier-token, punctuator-token, and data-value).

Hence, we consider regarding an embedded language statement as a stream of labeled tokens to be a suitable level of abstraction which allows us to precisely record the particularities of the embedded syntax. Consequently, the ELET is a container which holds a flat stream of labeled language tokens. See Listing 2 for an exemplified representation of Sec. 1.1's SQL.

```
1  $sql = {select-token, meta-char(*), from-token, tablename-token(Users),
2          where-token, fieldname-token(Passwd), metachar(=), metachar('),
3          stringliteral(mycatiscalled), metachar(')}
```

Listing 2. ELET-internal representation of Fig. 1.b's SQL statement

Identification of language elements: An analysis of common embedded languages, such as HTML, SQL, or JavaScript, along with an investigation of the current practice of code assembly yields the following observation: The set of existing token-types can be partitioned into three classes - static tokens, identifier tokens, and data literals (see Fig. 3):

Static tokens are statically defined terminals of the embedded language's grammar. Static tokens are either language keywords (such as SQL instructions, HTML tag- and attribute-names, or reserved JavaScript keywords) or meta-characters which carry syntactical significance (such as * in SQL, < in HTML, or ; in JavaScript).

Identifier tokens have values that are not statically defined within the language's grammar. However, they represent semi-static elements of the embedded code, such as names of SQL tables or JavaScript functions. As already observed and discussed, e.g., in [2] or [6], such identifiers are statically defined at compile-time of the hosting code. I.e., the names of used JavaScript functions are not dynamically created at run-time – they are hard-coded through constant string-literals in the hosting application.

Data literals represent values which are used by the embedded code. E.g., in Fig. 1.b the term "mycatiscalled" is such a literal. As it can be obtained from Fig. 1.a the exact value of this literal can change every time the hosting code is executed. Hence, unlike static and identifier tokens the value of data literals are dynamic at run-time.

Rules for secure assembly of embedded syntax: As previously stated, our methodology relies on *explicit* code creation. This means, for instance, the only way for a developer to add a static token representing a **select**-keyword to a

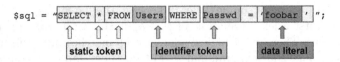

Fig. 3. Examples of the identified token classes

SQL-ELET is by explicitly stating his intent in the hosting source code. Consequently, the ELET has to expose a set of disjunct interfaces for adding members of each of the three token-classes, e.g. implemented via corresponding APIs. For the remainder of this paper, we assume that these interfaces are in fact implemented via API-calls. Please note: This is not a defining characteristic of our approach but merely an implementation variant which enables easy integration of the ELET into existing languages.

We can sufficiently model the three token-classes by designing the ELET's API according to characteristics identified above regarding the define-time of the tokens' values (see also Listing 3):

- API calls to add a static token to the ELET container are themselves completely static. This means, they do not take any arguments. Hence, for every element of this token-class, a separate API call has to exist. For instance, to add a `select`-keyword-token to a SQL-ELET a corresponding, dedicated `addSelectToken()`-method would have to be invoked.
- Identifier tokens have values which are not predefined in the embedded language's grammar. Hence, the corresponding ELET interface has to take a parameter that contains the token's value. However, to mirror fact that the values of such tokens are statically defined at compile-time and do not change during execution, it has to be enforced that the value is a constant literal. Furthermore, the embedded language's grammar defines certain syntactic restrictions for the token's value, e.g., the names of JavaScript variables can only be composed using a limited set of allowed characters excluding whitespace. As the arguments for the corresponding API methods only accept constant values these restrictions can be checked and enforced at compile-time.
- The values of data literals can be defined completely dynamically at runtime. Hence, the corresponding API methods do not enforce compile-time restrictions on the values of their arguments (which in general consist of string or numeric values).

```
1 sqlELET.addSelectToken();              // no argument
2 [...]
3 sqlELET.addIdentifierToken("Users");   // constant argument
4 ...
5 sqlELET.addStringLiteral(password);    // dynamic argument
6 [...]
```

Listing 3. Exemplified API-based assembly of Fig. 1.b's SQL statement

Through this approach, we achieve reliable protection against string-based code injection vulnerabilities: All untrusted, attacker-controlled values enter the application as either string or numeric values, e.g., through HTTP parameters or request headers. In order to execute a successful code injection attack the adversary has to cause the hosting application to insert parts of his malicious values into the ELET in a syntactical *code* context, i.e., as static or identifier tokens. However, the ELET interfaces which add these token types to the container do not accept dynamic arguments. Only *data* literals can be added this way. Consequently, a developer is simply not able to write source code which involuntarily and implicitly creates embedded *code* from attacker-provided string values. Hence, string-based code injection vulnerabilities are impossible as the ELET maintains at all times strict separation between the embedded *code* (the static and identifier tokens) and *data* (the data literal tokens).

2.4   Language Integration

While the outlined API-based approach fully satisfies our security requirements, it is a clear violation of our developer-oriented objective to preserve the embedded language's syntactic nature (see Sec. 2.1). A direct usage of the ELET API in a non-trivial application would result in source code which is very hard to read and maintain. Consequently, more suitable methods to add embedded syntax to the ELET have to be investigated.

A direct integration of the embedded syntax into the hosting language, for instance through a combined grammar, would be the ideal solution (see Listing 4). In such a case, the developer can directly write the embedded syntax into the hosting application's source code. First steps in this direction have been taken by Meijer et al. with Xen [14] and LINQ [13] which promote a subset of SQL and XML to be first class members of $C^\#$ However, this approach requires significant, non-trivial changes in the hosting language which are not always feasible. Also achieving full coverage of all features of the embedded language this way is still an open problem. Even advanced projects such as LINQ only cover a subset of the targeted languages and omit certain, sophisticated aspects which are hard to model within $C^\#$'s functionality.

```
1  sqlELET s = SELECT * FROM Users;
```

Listing 4. Direct integration of the embedded syntax

Instead, we propose a light-weight alternative which uses a source-to-source preprocessor: The embedded syntax is directly integrated into the hosting application's source code, unambiguously identified by predefined markers, e.g. $$ (see Listing 5). Before the application's source code is compiled (or interpreted), the preprocessor scans the source code for the syntactical markers. All embedded syntax that can be found this way is parsed accordingly to the embedded language's grammar and for each of the identified language tokens a corresponding API-call is added to the hosting source code (see Listing 6). The resulting API-based representation is never seen by the developer. He solely has to use unmodified embedded syntax. To add data values through hosting variables to the embedded syntax and offer capabilities for more dynamic code assembly (e.g., selective concatenation of syntax fragments) a simple meta-syntax is provided (see Sec. 3.2 for details).

```
1 | sqlELET s = $$ SELECT * FROM Users $$
```

Listing 5. Preprocessor-targeted integration of embedded syntax

```
1 | sqlELET s.addSelectToken().addStarMetachar().addFromToken()
2 |           .addIdentifierToken("Users");
```

Listing 6. Resulting API representation generated by the preprocessor

Compared to direct language integration, this approach has several advantages. For one, besides the adding of the ELET type, no additional changes to the original hosting language have to be introduced. The preprocessor is completely separate from the hosting language's parsing process. This allows flexible introduction of arbitrary embedded languages. Furthermore, achieving complete coverage of all features of the embedded language is not an issue. The only requirements are that the preprocessor covers the complete parsing process of the embedded language and the ELET provides interfaces for all existing tokens. Hence, covering obscure or sophisticated aspects of the embedded language is as easy as providing only a subset of the language.

However, our approach has also a drawback: The source code that is seen by the hosting application's parser differs from the code that the developer wrote. If parsing errors occur in source code regions which were altered by the preprocessor, the content of the parser's error message might refer to code that is unknown to the programmer. For this reason, a wrapped parsing process which examines the error messages for such occurrences is advisable.

2.5   External Interface

Finally, before communicating with the external entity (e.g., the database), the ELET-contained embedded syntax has to be serialized into a character-based representation. This is done within the external interface. From this point on, all meta-information regarding the tokens is lost. Therefore, the reassembly into a character-based representation has to be done with care to prevent the reintroduction of code injection issues at the last moment. The embedded syntax is provided within the ELET in a semi-parsed state. Hence, the external interface has exact and reliable information on the individual tokens and their *data/code*-semantics. Enforced by the ELET's restrictive interfaces, the only elements that may be controlled by the adversary are the *data* values represented by data literal tokens. As the exact context of these values within the embedded syntax is known to the external entity, it can reliably choose the correct encoding method to ensure the *data*-status of the values (see Sec. 3.3 for a concrete implementation).

2.6   Limitations

Before we document our practical implementation of our approach, we discuss the limitations of our methodology:

Run-time code evaluation: Certain embedded languages, such as JavaScript, provide means to create code from string values at run-time using function like `eval()`. Careless usage of such functionality can lead to string-based code injection vulnerabilities which are completely caused by the embedded code, such as DOM-based XSS [9]. The protection capabilities of our approach are limited to code assembly within the hosting language. To provide protection against this specific class of injection vulnerabilities, an ELET type has to be added to the embedded language and the `eval()`-functionality has to be adapted to expect ELET-arguments instead of string values.

Dynamically created identifier tokens: As explained in Sec. 2.3 the values of identifier tokens have to be defined through constant values. However, sometimes in real-life code one encounters identifiers which clearly have been created at run-time. One example are names of JavaScript variables which contain an enumeration component, such as `handler14`, `handler15`, etc. As this is not allowed by our approach, the developer has to choose an alternative solution like his purpose, e.g. arrays.

Permitting dynamic identifiers can introduce insecurities as it may lead to situations in which the adversary fully controls the value of an identifier token. Depending on the exact situation this might enable him to alter the control or data flow of the application, e.g. by exchanging function or variable names.

3   Practical Implementation and Evaluation

3.1   Choosing an Implementation Target

To verify the feasibility of our concepts, we designed and implemented our approach targeting the embedded languages HTML/JavaScript[2] and the hosting language Java. We chose this specific implementation target for various reasons: Foremost, XSS problems can be found very frequently, rendering this vulnerability-class to be one of the most pressing issues nowadays.Furthermore, as HTML and JavaScript are two independent languages with distinct syntaxes which are tightly mixed within the embedded code, designing a suiting preprocessor and ELET-type is interesting. Finally, reliably creating secure HTML-code is not trivial due to the lax rendering process employed by modern web browsers.

3.2   API and Preprocessor Design

The design of the ELET API for adding the embedded syntax as outlined in Sec. 2.3 was straightforward after assigning the embedded languages' elements

---

[2] NB: The handling of Cascading Style Sheets (CSS), which embody in fact a third embedded language, is left out in this paper for brevity reasons.

to the three distinct token types. Also, implementing the language preprocessor to translate the embedded code into API calls was in large parts uneventful. The main challenge was to deal with situations in which the syntactical context changed from HTML to JavaScript and vice versa. Our preprocessor contains two separate parsing units, one for each language. Whenever a HTML element is encountered that signals a change of the applicable language, the appropriate parser is chosen. Such HTML elements are either opening/closing `script`-tags or HTML attributes that carry JavaScript-code, e.g., event handlers like `onclick`.

As motivated in Sec. 2.1, we aim to mimic the string type's characteristics as closely as possible. For this purpose, we introduced a simple preprocessor meta-syntax which allows the programmer to add hosting values, such as Java strings, to the embedded syntax and to combine/extend ELET instances (see Listings 7 and 8). Furthermore, we implemented an iterator API which allows, e.g., to search a ELET's data literals or to split an ELET instance.

```
1  HTMLElet h $=$ Hello <b> $data(name)$ </b>, nice to see you! $$
```

Listing 7. Combination of embedded HTML with the Java variable **name**

3.3   Adding an External Interface for HTML Creation to J2EE

In order to assemble the final HTML output the external interface iterates through the ELET's elements and passes them to the output buffer. To deterministically avoid XSS vulnerabilities, all encountered data literals are encoded into a form which allows the browser to display their values correctly without accidentally treating them as *code*. Depending on the actual syntactical context a given element appears in an HTML page, a different encoding technique has to be employed:

- HTML: All meta-characters, such as <, >, =, " or ', that appear in data literals within an HTML context are encoded using HTML encoding ("&...;") to prevent the injection of rogue HTML tags or attributes.
- JavaScript-code: JavaScript provides the function `String.fromCharCode()` which translates a numerical representation into a corresponding character. To prevent code injection attacks through malicious string-values, all data literals within a syntactical JavaScript context are transformed into a representation consisting of concatenated `String.fromCharCode()`-calls.
- URLs: All URLs that appear in corresponding HTML attribute values are encoded using URL encoding ("%..").

In our J2EE based implementation the external interface's tasks are integrated into the application server's request/response handling. This is realized by employing J2EE's filter mechanism to wrap the `ServletResponse`-object. Through this wrapper-object servlets can obtain an `ELETPrinter`. This object in turn provides an interface which accepts instantiated ELETs as input (see Fig. 4.a). The serialized HTML-code is then passed to the original `ServletResponse`-object's output buffer. Only input received through the `ELETPrinter` is included in the final HTML-output. Any values that are passed to the output-stream through legacy character-based methods is logged and discarded. This way we ensure that only explicitly generated embedded syntax is sent to the users web browsers.

Implementing the abstraction layer in the form of a J2EE filter has several advantages. Foremost, no changes to the actual application-server have to be applied - all necessary components are part of a deployable application. Furthermore, to integrate our abstraction layer into an existing application only minor changes to the application's `web.xml` meta-file have to be applied (besides the source code changes that are discussed in Section 3.4).

3.4   Practical Evaluation

We successfully implemented a preprocessor, ELET-library, and abstraction layer for the J2EE application server framework. To verify our implementation's protection capabilities, we ran a list of well known XSS attacks [3] against a specifically crafted test application. For this purpose, we created a test-servlet that blindly echos user-provided data back into various HTML/JavaScript data- and code-contexts (see Fig. 8 for an excerpt).

```
1  protected void doGet(HttpServletRequest req, HttpServletResponse resp)
2      throws IOException
3  {
4      String bad = req.getParameter("bad");
5      [...]
6      HTMLelet h $=$ <h3>Protection test</h3> $$
7      h $+$ Text: $data(bad)$ <br /> $$
8      h $+$ Link: <a href="$data(bad)$">link</a> <br /> $$
9      h $+$ Script: <script>document.write($data(bad)$);</script><br /> $$
10     [...]
11     EletPrinter.write(resp, h);  // Writing the ELET to the output buffer
12     resp.getWriter().println(bad); // Is the legacy interface disabled?
13 }
```

Listing 8. Test-servlet for protection evaluation (excerpt)

Furthermore, to gain experiences on our approach's applicability in respect to non-trivial software we modified an existing software project to use our methodology. Porting an application to our approach requires to locate every portion of the application's source code which utilize strings to create embedded code. Such occurrences have to be changed to employ ELET semantics instead. Therefore, depending on the specific application, porting an existing code-base might prove to be difficult. We chose JSPWiki [5] as a porting target. JSPWiki is a mature J2EE based WikiWiki clone which was initially written by Janne Jalkanen and is released under the LGPL. More precisely, we chose version 2.4.103 of the software, a release which suffers from various disclosed XSS vulnerabilities [10]. The targeted version of the code consists of 365 java/jsp files which in total contain 69.712 lines of source code.

The porting process was surprisingly straightforward. JSPWiki's user-interface follows in most parts a rather clean version of the Model-View-Controller (MVC) pattern which aided the porting process. Besides the user-interface also the application's Wiki-markup parser and HTML-generator had to be adapted. It took a single programmer about a week to port the application's core functionality. In total 103 source files had to be modified.

As expected, all documented XSS vulnerabilities [10] did not occur in the resulting software. This resolving of the vulnerabilities was solely achieved by the porting process without specifically addressing the issues in particular.

Finally, to gain first insight of our approach's runtime behaviour, we examined the performance of the prototype. For this purpose, we benchmarked the unaltered JSP code against the adapted version utilizing the ELET paradigm. The performance tests were done using the HP LoadRunner tool, simulating an increasing number of concurrent users per test-run. The benchmarked applications were served by an Apache Tomcat 5.5.20.0 on a 2,8 GHz Pentium 4 computer running Windows XP Professional. In situations with medium to high server-load, we observed an overhead of maximal 25% in the application's response times (see Fig. 4.c). Considering that neither the ELET's nor the external interface's implementation have been specifically optimized in respect to performance, this first result is very promising. Besides streamlining the actual ELET implementation, further conceptual options to enhance the implementation's performance exist. For instance, integrating the abstraction layer directly into the application server, instead of introducing it via object wrappers would aid the overall performance.

4   Related Work

In the last decade extensive research concerning the prevention of code injection attacks has been conducted. In this section we present selected related approaches. In comparison to our proposed solution, most of the discussed techniques have the limitations that they are not centrally enforceable and/or are prone to false positives/negatives. Both characteristics do not apply to our technique.

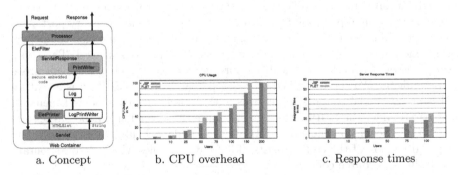

a. Concept   b. CPU overhead   c. Response times

Fig. 4. Implementation of the J2EE EletFilter

Manual protection and special domain solutions: The currently used strategy against XSS attacks is manually coded input filtering and output encoding. As long as unwanted HTML and JavaScript code is properly detected and stripped from all generated web pages, XSS attacks are impossible. However, implementing these techniques is a non-trivial and error prone task which cannot be enforced centrally, resulting in large quantities of XSS issues. In order to aid developers to identify XSS and related issues in their code, several information-flow based approaches for static source code analysis have been discussed, e.g., [4], [11], [7], or [23]. However, due to the undecidable nature of this class of problems such approaches suffer from false positives and/or false negatives.

Manual protection against SQL injection suffers from similar problems as observed with XSS. However, most SQL interpreters offer *prepared statements* which provide a secure method to outfit static SQL statements with dynamic data. While being a powerful migration tool to avoid SQL injection vulnerabilities, prepared statements are not completely bulletproof. As dynamic assembly of prepared statements is done using the string type, injection attacks are still possible at the time of the initial creation of the statement [22]. Furthermore, methods similar to prepared statements for most other embedded languages besides SQL do not exist yet. Therefore, dynamic assembly of embedded code in these languages has to employ similar mitigating strategies as mentioned in the context of XSS, with comparable results.

Type based protection: In concurrent and independent work [19], Robertson and Vigna propose a type based approach which is closely related to our ELET paradigm. They introduce dedicated datatypes that model language features of HTML and SQL. This way they enforce separation of the *content* and *structure* of the created syntax, two concepts that are similar to this paper's terms *data* and *code*.

However, their approach is not easily extensible to general purpose programming languages, such as JavaScript. Consequently, cross-site scripting caused by insecure dynamic assembly of JavaScript code (e.g., see Listing 8, line 9) is not prevented. In comparison, our approach is language independent and, thus, covers dynamically assembled JavaScript. Furthermore, notably absent from the paper is a discussion concerning the actual construction of the embedded code. For example, it is not explained, how the programmer creates HTML syntax in practice. As previously discussed, we consider such aspect to be crucial in regard to acceptance of the developer community.

Taint propagation: Run-time taint propagation is regarded to be a powerful tool for detecting string-based code injection vulnerabilities. Taint propagation tracks the flow of untrusted data through the application. All user-provided data is "tainted" until its state is explicitly set to be "untainted". This allows the detection if untrusted data is used in a security sensible context. Taint propagation was first introduced by Perl's taint mode. More recent works describe finer grained approaches towards dynamic taint propagation. These techniques allow the tracking of untrusted input on the basis of single characters. In independent concurrent works [16] and [18] proposed fine grained taint propagation to counter various classes of injection attacks. [2] describes a related approach ("positive tainting") which, unlike other proposals, is based on the tracking of trusted data. Furthermore, based on dynamic taint propagation, [21] describe an approach that utilizes specifically crafted grammars to deterministically identify code injection attempts. Finally, [24] proposes a fine grained taint mechanism that is implemented using a C-to-C source code translation technique. Their method detects a wide range of injection attacks in C programs and in languages which use interpreters that were written in C. To protect an interpreted application against injection attacks the application has to be executed by a recompiled interpreter.

Compared to our approach, dynamic taint propagation provides inferior protection capabilities. Taint-tracking aims to prevent the *exploitation* of injection vulnerabilities while their fundamental causes, string-based code assembly and the actual vulnerable code, remain unchanged. Therefore, in general the sanitazion of the tainted data still relies on string operations. The application has to "untaint" data after applying manually written validation and encoding function, a process which in practice has been proven to be non-trivial and error-prone. This holds especially true in situations where limited user-provided HTML is permitted. E.g., no taint-tracking solution would have prevented the myspace.com XSS that was exploited by the Samy-worm [8]. In our approach, even in cases where user-provided HTML is allowed, such markup has to be parsed from the user's data and recreated explicitly using ELET semantics, thus, effectively preventing the inclusion of any unwanted code. Furthermore, unlike our approach taint-tracking is susceptible to second-order code injection vulnerabilities [17] due to its necessary classification of data origins as either *trusted* or *untrusted*. In the case of second-order code injection the attacker is able to reroute his attack through a trusted component (e.g., temporary storage of an XSS attack in the DB).

Embedded language integration: Extensive work has been done in the domain of specifically integrating a certain given embedded language into hosting code. Especially SQL and XML-based languages have received a lot of attention. However, unlike our approach, the majority of these special purpose integration efforts neither can be extended to arbitrary embedded languages, nor have been designed to prevent code injection vulnerabilities. In the remainder of this section we describe selected publications that share similarities with our approach:

As previously discussed in Section 2.3, LINQ [13] promotes subsets of XML and SQL to be first class members of $C^\#$. Furthermore, E4X [20] is a related approach which integrates XML into JavaScript. The applied techniques are entirely data-centric with the goal to soften the object-relational impedance mismatch [14]. Due to this characteristic, their approach is not easily extensible to procedural embedded languages, such as JavaScript. In comparison, our approach is completely independent from the characteristics of the embedded language. [12] describes SQL DOM. A given database schema can be used to automatically generate a SQL Domain Object Model. This model is transformed to an API which encapsulates the capabilities of SQL in respect to the given schema, thus eliminating the need to generate SQL statements with the string datatype. As every schema bears a schema-specific domain object model and consequently a schema-specific API, every change in the schema requires a re-generation of the API. Finally, SQLJ [1] and Embedded SQL [15], two independently developed mechanisms to combine SQL statements either with Java or C respectively, employ a simple preprocessor. However, unlike our proposed approach these techniques only allow the inclusion of static SQL statements in the source code. The preprocessor then creates hosting code that immediately communicates the SQL code to the database. Thus, dynamic assembly and processing of embedded syntax, as it is provided in our proposed approach via the ELET, is not possible.

5   Conclusion

In this paper we proposed techniques to enhance programming languages with capabilities for secure and explicit creation of embedded code. The centerpiece of our syntax-assembly architecture is the ELET (see Section 2.3), an abstract datatype that allows the assembly and processing of embedded syntax while strictly enforcing the separation between *data* and *code* elements. This way injection vulnerabilities that are introduced by implicit, string-serialization based code-generation become impossible. To examine the feasibility of the proposed approach, we implemented an integration of the embedded languages HTML and JavaScript into the Java programming language. Usage of our approach results in a system in which every creation of embedded syntax is an explicit action. More specifically, the developer always has to define the exact particularities of the assembled syntax precisely. Therefore, accidental inclusion of adversary-provided semantics is not possible anymore. Furthermore, the only way to assemble embedded syntax is by ELET, which effectively prevents programmers from taking insecure shortcuts. A wide adoption of our proposed techniques would reduce the attack surface of code injection attacks significantly.

Acknowledgements

We wish to thank Erik Meijer for giving us valuable feedback and insight into LINQ.

References

1. American National Standard for Information Technology. ANSI/INCITS 331.1-1999 - Database Languages - SQLJ - Part 1: SQL Routines using the Java (TM) Programming Language. InterNational Committee for Information Technology Standards (formerly NCITS) (September 1999)

2. Halfond, W.G.J., Orso, A., Manolios, P.: Using positive tainting and syntax-aware evaluation to counter sql injection attacks. In: 14th ACM Symposium on the Foundations of Software Engineering, FSE (2006)
3. Hansen, R.: XSS (cross-site scripting) cheat sheet - esp: for filter evasion, http://ha.ckers.org/xss.html (05/05/07)
4. Huang, Y.-W., Yu, F., Hang, C., Tsai, C.-H., Lee, D.-T., Kuo, S.-Y.: Securing web application code by static analysis and runtime protection. In: Proceedings of the 13th conference on World Wide Web, pp. 40–52. ACM Press, New York (2004)
5. Jalkanen, J.: Jspwiki. [software], http://www.jspwiki.org/
6. Johns, M., Beyerlein, C.: SMask: Preventing Injection Attacks in Web Applications by Approximating Automatic Data/Code Separation. In: 22nd ACM Symposium on Applied Computing (SAC 2007) (March 2007)
7. Jovanovic, N., Kruegel, C., Kirda, E.: Pixy: A static analysis tool for detecting web application vulnerabilities. In: IEEE Symposium on Security and Privacy (May 2006)
8. Kamkar, S.: Technical explanation of the myspace worm (October 2005), http://namb.la/popular/tech.html (01/10/06)
9. Klein, A.: DOM Based Cross Site Scripting or XSS of the Third Kind (September 2005), http://www.webappsec.org/projects/articles/071105.shtml (05/05/07)
10. Kratzer, J.: Jspwiki multiple vulnerabilitie. Posting to the Bugtraq mailinglist (September 2007), http://seclists.org/bugtraq/2007/Sep/0324.html
11. Livshits, B., Lam, M.S.: Finding security vulnerabilities in java applications using static analysis. In: Proceedings of the 14th USENIX Security Symposium (August 2005)
12. McClure, R.A., Krueger, I.H.: Sql dom: compile time checking of dynamic sql statements. In: Proceedings of the 27th International Conference on Software Engineering (2005)
13. Meijer, E., Beckman, B., Bierman, G.: LINQ: Reconciling Objects, Relations, and XML In the.NET Framework. In: SIGMOD 2006 Industrial Track (2006)
14. Meijer, E., Schulte, W., Bierman, G.: Unifying tables, objects, and documents. In: Declarative Programming in the Context of OO Languages (DP-COOL 2003), vol. 27. John von Neumann Institute of Computing (2003)
15. MSDN. Embedded sql for c, http://msdn.microsoft.com/library/default.asp?url=/library/en-us/esqlforc/ec_6_epr_01_3m03.asp (27/02/07)
16. Nguyen-Tuong, A., Guarnieri, S., Greene, D., Shirley, J., Evans, D.: Automatically hardening web applications using precise tainting. In: 20th IFIP International Information Security Conference (May 2005)
17. Ollmann, G.: Second-order code injection. Whitepaper, NGSSoftware Insight Security Research (2004), http://www.ngsconsulting.com/papers/SecondOrderCodeInjection.pdf
18. Pietraszek, T., Berghe, C.V.: Defending against injection attacks through context-sensitive string evaluation. In: Valdes, A., Zamboni, D. (eds.) RAID 2005. LNCS, vol. 3858, pp. 124–145. Springer, Heidelberg (2006)
19. Robertson, W., Vigna, G.: Static Enforcement of Web Application Integrity Through Strong Typing. In: USENIX Security (August 2009)
20. Schneider, J., Yu, R., Dyer, J. (eds.): Ecmascript for xml (e4x) specification. ECMA Standard 357, 2nd edn. (December 2005), http://www.ecma-international.org/publications/standards/Ecma-357.htm

21. Su, Z., Wassermann, G.: The essence of command injection attacks in web applications. In: Proceedings of POPL 2006 (January 2006)
22. von Stuppe, S.: Dealing with sql injection (part i) (February 2009), http://sylvanvonstuppe.blogspot.com/2009/02/ dealing-with-sql-injection-part-i.html (04/24/09)
23. Wassermann, G., Su, Z.: Static detection of cross-site scripting vulnerabilities. In: Proceedings of the 30th International Conference on Software Engineering, Leipzig, Germany, May 2008. ACM Press, New York (2008)
24. Xu, W., Bhatkar, S., Sekar, R.: Taint-enhanced policy enforcement: A practical approach to defeat a wide range of attacks. In: 15th USENIX Security Symposium (August 2006)

Idea: Reusability of Threat Models – Two Approaches with an Experimental Evaluation

Per Håkon Meland, Inger Anne Tøndel, and Jostein Jensen

SINTEF ICT, SP Andersens vei 15 B, N-7465 Trondheim, Norway
{per.h.meland,inger.a.tondel,jostein.jensen}@sintef.no
http://www.sintef.com/

Abstract. To support software developers in addressing security, we encourage to take advantage of reusable threat models for knowledge sharing and to achieve a general increase in efficiency and quality. This paper presents a controlled experiment with a qualitative evaluation of two approaches supporting threat modelling - reuse of categorised misuse case stubs and reuse of full misuse case diagrams. In both approaches, misuse case threats were coupled with attack trees to give more insight on the attack techniques and how to mitigate them through security use cases. Seven professional software developers from two European software companies took part in the experiment. Participants were able to identify threats and mitigations they would not have identified otherwise. They also reported that both approaches were easy to learn, seemed to improve productivity and that using them were likely to improve their own skills and confidence in the results.

1 Introduction

For a time now, we have seen a trend towards more connected, dynamic and complex software systems, along with an increased number of threats and examples of cybercrime. According to Torr [3], "... even security-conscious products can fall prey when designers fail to understand the threats their software faces or the ways in which adversaries might try to attack it." Torr suggests including threat modelling as a software development activity to get an overview of potential threats towards applications and how to resolve these. The term *threat models* is typically used for models like attack trees [4] and threat trees [5], but also data flow diagrams [3] and misuse case diagrams [6]. The threat model types mentioned above all utilise well known structures to developers, like trees and UML. It is also possible to utilise more formal model types as a basis for threat modelling, like e.g. Jackson's problem diagrams [7] or an extension to the agent-oriented requirements modelling language i* [8]. We agree that threat modelling is a good approach to address security problems during development. However, in our experience threat modelling can be a time consuming (i.e. expensive) and complex task to perform for software developers without extensive security expertise. Also, we have seen that applications with similar characteristics tend to end up with very similar threat models. The work presented in

F. Massacci, D. Wallach, and N. Zannone (Eds.): ESSoS 2010, LNCS 5965, pp. 114–122, 2010.

this paper aims at supporting threat modelling activities and thereby contribute to improving software security aspects in general. By providing developers with reusable threat models, security experts can share their knowledge and lessons learned from the past, and we believe that the modelling efficiency and quality will be substantially improved. In our work we are primarily concerned with supporting developers that are not necessarily security experts. We thus focus on model types that are easy to understand and likely to be familiar to developers. Earlier experiments indicate that misuse cases and attack trees are easy to grasp [9]. We have therefore decided to use these as the basis for our research. We have conducted an experiment to help us evaluate the form of reusable threat models. A set of developers from two European software development companies created misuse case diagrams for two experiment case studies, but for each used a slightly different method of reusing security knowledge from threat models.

We are not aware of any work on reusability of misuse case diagrams. Reuse of textual misuse cases and security use cases have already been proposed and described by Sindre et al. [10]. Regarding attack trees, reusability was pointed out by Schneier as a main advantage already in 1999 [4], but we have not seen much of that in practice. We are familiar of two experiments evaluating misuse cases and attack trees together. Diallo et al. [11] describes a limited experiment where two researchers used the different model types on one single example application, and documented their experiences. Opdahl and Sindre [9] describes an experimental evaluation that included in total 63 students. Performance and perception of the two different model types were evaluated. Results from both experiments suggest that misuse cases and attack trees would probably benefit from being used in combination as they have different strengths. Results from the experimental evaluation also showed that the students tended to prefer misuse case diagrams over the textual descriptions.

2   Research Method

We have been working with two different approaches to reuse of misuse cases:

1. **Reuse of misuse case stubs.** Importing generic threats in the form of stand-alone misuse case elements, which are connected to security use cases and attack trees.
2. **Reuse of full misuse case diagrams for given application types.** Representative misuse case diagrams for specific domains and application types are used as a source for the application at hand.

Both approaches are based on the claim that recognising relevant model items from organised lists or examples is easier than creating something from scratch. In order to get more knowledge on applicability for ordinary developers, we performed an experiment specifically addressing and comparing these approaches.

As we only had a limited number of candidates for the experiment, we chose a Latin-Squares experimental design. This way all participants could construct threat models using both approaches, while we could control the order in which the different approaches were used. To summarise, experiment participants were divided into two groups working with case A, where group I started out with approach 1 (M1), while group II started out with approach 2 (M2). Then both groups got a new case (case B) and now performed threat modelling with the other approach (group I used approach 2 (M2) while group II used approach 1 (M1)). Before the experiment started, all participants filled out a questionnaire on their background. Then after the use of each method, they filled out a questionnaire related to perception based on the Technology Acceptance Model (TAM) [12].

The following material was used during the experiment:

- An **experiment manual**, containing a description of both methods (M1 and M2), along with both case study descriptions.
- **Questionnaires** on background information, completed before the experiment began, and two post-method questionnaires.
- An **"attack tree forest"**, a document and a set of model files containing 30 attack trees for typical software attacks (used for both M1 and M2).
- A **catalogue of misuse case diagrams**, a document and a set of model files containing 8 full misuse case diagrams for various applications. Misuse case threats were associated to attack trees and mitigating security activities. Figure 1 shows a misuse case diagram example for a typical Web shop application.
- A **catalogue of threats and misuse case stubs**, a document organised according to 7 threat categories[1], where each category had an average of about 6 threat misuse case stubs associated to it. Misuse case threats were associated to attack trees and mitigating security activities. Figure 2 shows examples of such stubs within the category *Information disclosure.*
- A **guide** explaining the main principles on how to perform threat modelling using misuse cases and attack trees (with examples).
- The **SeaMonster**[2] tool, supporting both misuse case and attack tree diagrams. The participants were encouraged to use this tool, but this was not a requirement.

All participation was voluntary and anonymous, and most participants were familiar with practical threat modelling from exercises conducted earlier. Both the guide and the SeaMonster tool were made available one month in advance of the experiment. The participants were divided into two groups (Group I or Group II), which identified the method they were to start with. Both groups started with case study A, and spent about 2-3 hours on this before they handed in their model and text files to the experiment leader, and filled out the post-method questionnaire. A similar amount of time was then spent on case study

---

[1] We chose to use the STRIDE (*Spoofing, Tampering, Repudiation, Information disclosure, Denial of service and Elevation of privileges* categories [5], with the addition of a category *Other* for threats difficult to place.

[2] Described by Meland et al. [13] and available from http://sourceforge.net/projects/seamonster/

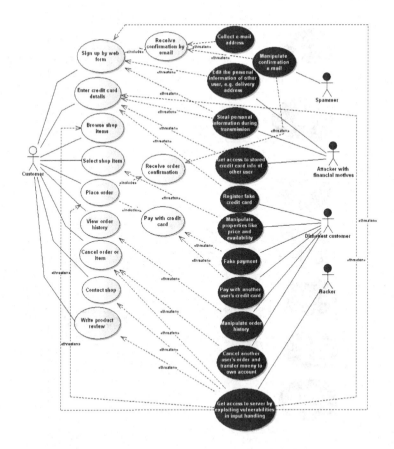

Fig. 1. Example of a *Web shop* misuse case diagram used with M2 (attack tree relations omitted)

B with the other method, before filling out another post-method questionnaire. Both methods started with a common part, which was to analyse the case study description assisted by a set of steps defined in the guide. Based on the identified actors, assets and functionality, the participants would then create the actual misuse case diagrams by applying one of these approaches:

M1: Using existing misuse cases as a source for creating your own misuse cases: *When you are creating the misuse case(s), go through the "misuse case diagram catalogue" and look for similar applications and diagram elements that are also relevant for you. Modify the diagram or add elements to you own misuse case model.* Of course, the catalogue did deliberately not contain any perfect matches for any of the case studies, and it was composed from the documentation of several development projects and examples found in the literature.

M2: Using categorised threats and misuse case stubs as a source for creating your own misuse cases: *When you are creating the misuse case(s),*

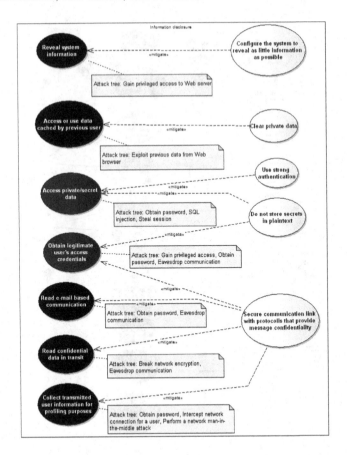

Fig. 2. Example of misuse case stubs used with M1 (with attack tree relations)

go through the "Threat and misuse case stub catalogue" and look for relevant threats and misuse case stubs. Select and add these to you own misuse case model.

We chose to present the participants with two somewhat different application areas, and created descriptions that would be typical outcome of an initial customer meeting. The full descriptions were 1-2 pages for each case study.

Case study A: Online store for mobile phones: This case study was loosely based on the *mobile Web shop* case previously used by Opdahl et al. [9]. The context was an online store with digital content like music, videos, movies, ring tones, software and other products for mobile equipment like e.g. 3G mobile phones. It was mainly to be accessed by customers directly through their mobile phones, but other than that resembles any other online store.

Case study B: Electronic Patient Record (EPR) system: This case study was created in order to have a different environment, stakeholders and assets compared to case study A. An electronic patient record system (EPR) is used

Table 1. Overview of threat types for M1/M2 and case A/B, and used threat categories

	Threats identified per participant				Used threat categories
	Case A	Case A	Case B	Case B	M1 + M2
	M1	**M2**	**M1**	**M2**	
Spoofing	1,5	3	3,5	3	5 of 5
Tampering	2,5	2	1,5	3	4 of 10
Repudiation	0	0	0,5	0	1 of 2
Information disclosure	2	2	4	1	5 of 8
Denial of Service	0	1	0,5	1	1 of 1
Elevation of privileges	1	2	1,5	2	4 of 6
Other	1	0	0,5	1	3 of 12

by clinicians to register and share information within and between hospitals. Much of the patient information should be regarded as strictly private/sensitive, and one fundamental assumption is that the system is only available on a closed health network for hospitals (not accessible from the Internet or terminals outside of the hospitals).

3   Results and Discussion

Participant background: In total seven participants took part in the experiment: five from Company A (CA) and two from Company B (CB). The group was heterogeneous, meaning that the participants' background differed when it came to experience in software security, misuse cases, attack trees, prior knowledge of the guide explaining threat modelling and the SeaMonster tool. In general, attack trees were less familiar to the participants compared to misuse cases.

Coverage of results: To determine similarity regarding the number of and types of threats, we analysed the misuse cases identified by the participants. As the threats in the catalogues for M1 and M2 were not always phrased the same, we manually grouped similar threats. We continued to use the STRIDE + *Other* categories for the analysis, and table 1 gives an overview of the threats types according to method and case. We can not see any significant differences based on this small number of participants. The variations are more likely to be because of differing focus of the participants than because of the approach used. As an example, more threats of the type *Spoofing* were identified for M2 than for M1 in case A, but in case B this was opposite.

Effectiveness/performance: For case A (online store for mobile phones) there was an existing misuse case diagram for a "Webshop" that provided a pretty good match, and most participants used this as a starting point. For case B (EPR system) it was not so obvious which existing misuse case diagrams were relevant, and also harder to analyse which diagrams have been considered for reuse. The participants indicated more domain experience with case A than case B. Regarding the threat stubs, it seems that some threat categories have been more extensively used than others. An overview of this is shown in the rightmost column of table 1. Typically the threats that were not used were application specific. An obvious observation is that the category *Other* has not been extensively used, and may have a tendency to be overlooked. Threats currently placed here should be moved to other categories if possible.

Perception: Perception was measured by the participants filling out a questionnaire on their experiences with the different approaches to threat model reuse. In the analysis work we grouped the questions asked according to subject, and addressed the participants' responses regarding the different issues. The differences in mean responses is not large enough to account for deviation in the answers of individual participants. Therefore, we see these results as more of indications than proof.

Identification of threats and countermeasures: This method helped me find many threats I would never have identified otherwise. I was able to identify many threats in addition to the ones suggested. This method helped me find many security use cases I would never have identified otherwise. I was able to identify many additional security use cases as well. For all methods, participants in general found that they were able to identify threats and security use cases they would not have identified otherwise. This supports that providing reusable threat models is a good idea, and that including countermeasures in the form of security use cases is helpful to developers. There were no major differences between the methods, except that for M1 participants experienced with the domain were more likely to report that they could identify additional threats than those with less experience.

Attack trees: It was easy to find relevant attack trees using this method. The attack trees that were suggested were mostly irrelevant. I hardly used the provided attack tree forest. The attack tree forest seemed pretty complete. For all methods, participants in general found that they were able to identify relevant attack trees. Participants did not agree on the relevance and usefulness of the attack trees and the completeness of the attack tree forest. In the experiment, the participants were not expected to create new or update existing attack trees, but to analyse the attacks to identify mitigating security use cases.

Usability: This method was easy to learn. This method made me spend little time creating the misuse cases. I feel more productive when using this method. It would be easy for me to become more skillful using this method. Using this method is a good idea. I think there are better ways of creating misuse cases. I would like to use this method in the future. The participants found the methods easy to learn, while M2 scored slightly better on the time spent. For M2 at Company A the participants did not create misuse case diagrams that included use cases (functionality), therefore the resulting diagrams contained less information compared to M1 and M2 at Company B. For both M1 and M2 the participants felt more productive using the method, but neutral to time consumption. Generally, they said they would become more skillful using the methods and that it was a good idea to use these methods, while neutral or agreed to future use. When asked about better ways of creating misuse cases, the participants were in general neutral (though some agreed). We received a comment that M2 seemed to be more "straight forward" and that it made the reflection more "application centric" compared to M1, which required finding similarities between different systems.

Result: I am confident in the results I have created. Using this method would enhance the quality when creating misuse cases. Participants were generally confident in the results they created. For M1 and for M2 as performed by Company B, participants also reported that they think using the method will enhance quality. M2 at Company A scored lower, possibly because their models did not include functionality.

Tool support: Proper tool support makes this method easier. Proper tool support makes this method more efficient. Tool support was considered important by some, but far from all participants. This is a bit surprising as we have viewed tool support as an important prerequisite for successful reuse of threat models. This result may however be influenced by the current version of SeaMonster, which is a prototype with several known limitations. It was not our goal to evaluate this specific tool, but to get opinions on tool support in general.

Quality of catalogue: I hardly used the provided security knowledge in the catalogue. The catalogue for this method seemed pretty complete. Participants reported that they used the provided security knowledge to a different extent. They also did not think the same on the completeness of the catalogue. In general, the participants that had experience with software security and threat modelling were more likely to report on weaknesses in the provided material.

4 Conclusion and Further Work

We have performed a controlled experiment with the aim of evaluating two different methods to threat model reuse. We could not conclude that one approach is superior to the other, and there are no significant differences in the types and number of threats identified and in user perception. However, the results and feedback from the participants indicates that reuse itself will improve threat modelling.

- For both methods, participants generally found that they were able to identify threats and security use cases they would not have identified otherwise.
- Regardless of method, participants generally found that they were able to identify relevant attack trees.
- Responses indicate that, for both methods, it is easy to learn how to use them and using them seems to improve productivity.
- Independently of the method, participants are in general confident in the results they have created.
- Participants generally got the impression that using these methods they would become more skillful and they are good ideas.

We plan to perform further experiments with larger groups in order to obtain more statistically valid results, and also work towards improvements of our current approaches to reuse. The reusable material will be available to security experts and software developers through an on-line security model repository [14].

Acknowledgment

The research leading to these results has received funding from the European Community Seventh Framework Programme (FP7/2007-2013) under grant agreement no 215995.

References

1. McGraw, G.: Software Security: Building Security In. Addison-Wesley, Reading (2006)
2. Mouratidis, H., Giorgini, P., Manson, G.: When security meets software engineering: a case of modelling secure information systems. Information Systems 30(8), 609–629 (2005)
3. Torr, P.: Demystifying the threat modeling process. IEEE Security & Privacy 3(5), 66–70 (2005)
4. Schneier, B.: Attack trees. Dr. Dobb's Journal (1999)
5. Swiderski, F., Snyder, W.: Threat modeling. Microsoft Press, Redmond (2004)
6. Sindre, G., Opdahl, A.L.: Eliciting security requirements with misuse cases. Requirements Engineering 10(1), 34–44 (2005)
7. Haley, C., Laney, R., Moffett, J., Nuseibeh, B.: Security requirements engineering: A framework for representation and analysis. IEEE Transactions on Software Engineering 34(1), 133–153 (2008)
8. Liu, L., Yu, E., Mylopoulos, J.: Security and privacy requirements analysis within a social setting. In: IEEE International Conference on Requirements Engineering, pp. 151–161 (2003)
9. Opdahl, A.L., Sindre, G.: Experimental comparison of attack trees and misuse cases for security threat identification. Information and Software Technology 51(5), 916–932 (2009)
10. Sindre, G., Firesmith, D.G., Opdahl, A.L.: A reuse-based approach to determining security requirements. In: Proceedings of the 9th international workshop on requirements engineering: foundation for software quality, REFSQ 2003 (2003)
11. Diallo, M.H., Romero-Mariona, J., Sim, S.E., Alspaugh, T.A., Richardson, D.J.: A comparative evaluation of three approaches to specifying security requirements. In: 12th Working Conference on Requirements Engineering: foundation for Software Quality, REFSQ 2006 (2006)
12. Davis, F.: Perceived usefulness, perceived ease of use, and user acceptance of information technologies. MIS Quarterly 13(3), 319–340 (1989)
13. Meland, P.H., Spampinato, D.G., Hagen, E., Baadshaug, E.T., Krister, K.M., Vell, K.S.: SeaMonster: Providing tool support for security modeling. In: Norsk informasjonssikkerhetskonferanse, NISK 2008, Tapir (2008)
14. Meland, P.H., Ardi, S., Jensen, J., Rios, E., Sanchez, T., Shahmehri, N., Tøndel, I.A.: An architectural foundation for security model sharing and reuse. In: Proceedings of the Fourth International Conference on Availability, Reliability and Security (ARES2009), pp. 823–828. IEEE Computer Society, Los Alamitos (2009)

Model-Driven Security Policy Deployment: Property Oriented Approach

Stere Preda, Nora Cuppens-Boulahia, Frédéric Cuppens,
Joaquin Garcia-Alfaro, and Laurent Toutain

IT TELECOM Bretagne CS 17607, 35576 Cesson-Sévigné, France
{first_name.surname}@telecom-bretagne.eu

Abstract. We address the issue of formally validating the deployment of access control security policies. We show how the use of a formal expression of the security requirements, related to a given system, ensures the deployment of an anomaly free abstract security policy. We also describe how to develop appropriate algorithms by using a theorem proving approach with a modeling language allowing the specification of the system, of the link between the system and the policy, and of certain target security properties. The result is a set of proved algorithms that constitute the certified technique for a reliable security policy deployment.

1 Introduction

Security is concerned with assets protection. Securing the access to a file server, guaranteeing a certain level of protection of a network channel, executing particular counter measures when attacks are detected, are appropriate examples of security requirements for an information system. Such security requirements belong to a guide usually called the access control security policy. Deploying the policy means enforcing (i.e., configuring) those security components or mechanisms so that the system behavior is the one specified by the policy. Successfully deploying the policy depends not only on the complexity of the security requirements but also on the complexity of the system in terms of architecture and security functionalities.

Specifying, deploying and managing the access control rules of an information system are some of the major concerns of security administrators. Their task can be simplified if automatic mechanisms are provided to distribute or update, in short to deploy the policy in complex systems. A common approach is the formalization of the security policy, based on an access control model and the application of a *downward* process (i.e., the translation and refinement of the formal policy requirements into concrete security device configurations). Though PBNM (*Policy Based Network Management*) architectures (cf. [rfc 3198]) are such examples, a challenging problem persists: proving the deployment process to be *correct* with respect to some initial target security properties and ensuring that no ambiguities (e.g., inconsistencies [13]) are added within this process.

In this paper we aim at establishing a formal frame for the deployment of security policies in information systems. We formally prove the process of deploying

F. Massacci, D. Wallach, and N. Zannone (Eds.): ESSoS 2010, LNCS 5965, pp. 123–139, 2010.

a security policy related to an information system. To do so, we require (1) an expressive security access control model that covers a large diversity of security requirements, (2) a modeling language for system specification and (3) a formal expression of security properties modeling the relationships between the security policy and the system it was designed for. We propose a formal technique that combines the use of access control policies expressed in the OrBAC (Organization-Based Access Control) language [1] together with specifications based on the B-Method [2]. Our proposal avoids, moreover, the existence of inconsistencies derived from the deployment [13].

Paper Organization — Section 2 gives the motivation of our work and some related works. Section 3 presents the model on which we base our approach and establishes some prerequisites necessary for our proposal. Section 4 formally defines the link between a policy and a system, including the expression of some security properties. Section 5 provides a discussion upon our approach.

2   Motivation and Related Work

The policy-based configuration of security devices is a cumbersome task. Manual configuration is sometimes unacceptable: the security administrator's task becomes not only more difficult but also error-prone given the anomalies he/she may introduce. Guaranteeing the deployment of anomaly-free configurations in complex systems is achievable if the policy is first formalized based on an access control model and then automatically translated into packages of rules for each security device. This is the current approach in PBNM architectures where the PDP, *Policy Decision Point*, is the intelligent entity in the system and the PEPs, *Policy Enforcement Points*, enforce its decisions along with specific network protocols (e.g., Netconf [rfc 4741]). Obtaining these packages of rules (i.e., the configurations of PEPs) is the result of the *downward* translation process: the abstract policy, given the system architecture, is compiled through a set of algorithms at the PDP level into, for example, firewall scripts and IPsec tunnel configurations — all the way through bearing the system architecture details (interconnections and capabilities) [19].

The correctness of these algorithms is a crucial aspect since the system configuration must reflect the abstract policy. One can simply design such algorithms using imperative languages and then validate them via specific tools [18]. Imperative program verification is performed in three steps: (1) the specification of a program is first formalized by some properties based on a first order logic; (2) an automatic process for analyzing the program extracts its semantics, i.e., a set of equations that define the program theory in the first order logic; (3) a proof system is finally used to prove that the program has the properties of step (1) given the extracted data in step (2). The main concerns with such approaches are: (a) the verification of these properties is realized at the end and not during the algorithm implementation; (b) it is difficult to express the *interesting* properties: not only those concerning the design of operations which may be seen as generic (e.g., termination of loops or lack of side effects), but also those reflecting, in our case, some network security aspects (e.g., no security *anomalies* are introduced during the deployment of the policies [13]). To cope with these issues, we claim that the algorithms intended for enforcing deployment of policies have to be designed with proof-based development methods, like the B-Method, to allow the expression and verification of important properties (e.g., security properties).

Formal validation or security policy deployment has already been addressed in the literature. The approaches in [15] and [16] seem to be the closest to ours. Jürjens et al. propose in [15] to apply UMLsec [14] to analyze some security mechanisms enforced with respect to a security policy. UMLsec is an extension of UML which allows the expression of some security-related information in UML diagrams. Stereotypes and tags are used to formulate the security requirements. Analyzing the mechanisms means verifying whether the requirements are met by the system design. For this purpose two models are proposed: (1) a Security Requirements Model which includes architectural or behavioral system details in a prescriptive manner and (2) a Concretized Model summarizing a concrete architecture which should satisfy the security requirements. Both models appear as UMLsec diagrams. The verification is realized using the UMLsec tools which includes several plugins that uses (1) SPIN (*Simple Promela Interpreter*) for model-checking and (2) SPAAS (an *automated theorem prover for first-order logic with equality*) or Prolog as theorem provers given that some UMLsec sequence diagrams are automatically translated to first-order formulas. Applying the right plugin depends on the scenario, i.e., the architecture and the security requirements. However, the approach in [15] presents some drawbacks. First, no abstract model is employed for modeling the policy. Second, the Concretized Model must already exist in order to automatically derive the first-order formulas to be automatically proved by SPAAS. And finally, the expressions of security properties are application-dependent: there are no generic properties dealing with the anomalies that could exist within single- or multi-component network security policies.

Laborde et al. present in [16] a different solution to the problem of deploying security policies: the use of (1) Petri Nets as the language to specify the system and (2) CTL (*Computational Tree Logic*) as the language to express the security properties. Four generic system *functionalities* are identified and modeled as different Petri Nets which are then interconnected in order to specify the system (i.e., each security device is modeled by a Petri Net): channel (e.g., network links), transformation (IPsec and NAT), filtering (to include the firewalls) and the *end-flow* functionality for the hosts (the active and passive entities in the network). The Petri Nets transitions for each PEPs (here, firewalls and IPsec tunnels) are guards which actually represent the security rules to be enforced by the PEP. The policy model is RBAC-based. However, no clear downward (i.e., refinement) approach is defined and no algorithms for selecting the right PEP are described either. The model-checking verification is realized after manually deploying the policy and consequently this approach can be applied to relatively simple architectures. Besides, it is not clear whether other functionalities (e.g., intrusion detection performed by IDSs) can be taken into account with one of the four functionalities.

The research presented in [22] proposes the use of Event B specifications to provide a link between the two levels of abstractions provided by the OrBAC model. As further research perspectives, the authors mentioned that their models can be reused for further developments of real infrastructures with respect to their security policy. We consider our work as a natural continuation of such a research line. As we did, the authors in [22] chose not to address the proof of an OrBAC policy in terms of conflicts. The use of automatic tools (e.g., MotOrBAC [4]), allows us to assume that the policy is consistent and free of anomalies. Similarly, Coq was proposed in [7] to derive OCaml algorithms for conflict detection in firewall policies. The use of Coq as a theorem prover to derive refinement algorithms can also be found in [21]. The authors provide a solution to detect and remove conflicts in policies defined as tuples <permission/prohibition, subject, read/write/execute, object>. Finally, and regarding the security policy deployment domain, there exist in the literature several proposals. Some are more or less RBAC-based (e.g., proposal presented in [5]), others propose different languages for the high level policy definition. Although the efforts are significant [10], often such languages are not generic enough [17], covering only some specific security applications (e.g., host firewalls, system calls management); or they do not address some key policy matters like the conflict management or the dynamic and contextual security requirements [8].

3   Model and Notation

We propose a refinement process that guarantees anomaly-free configurations ([13]). The process derives a global policy into specific configurations for each security component in the system. Our proposal provides the set of algorithms for such a refinement process, and proves the correctness of the outgoing algorithms. We briefly justify in the sequel the choice of our formalisms. We also describe the necessary concepts to establish the link between policies and architectures.

3.1   Choice of OrBAC and B-Method

The OrBAC [1] (Organization-Based Access Control) model is an extended RBAC [20] access control model which provides means to specify contextual security requirements. It allows the expression of a wide range of different requirements both static and contextual. OrBAC is well-known for being a robust language for defining complex administration tasks at the organizational level. Existing automatic tools (e.g., MotOrBAC [4]) ease, moreover, the administration of tasks using this model.

In contrast to model checking, we choose the B-Method – a theorem proving approach – for various reasons. First, the performances of theorem-proving tools are not influenced by system complexity. For example, we do not make any assumption concerning the number of nodes in the architecture. Second, the B Method eases the use of refinement paradigms. Even if *B refinement* does not necessarily mean an enhancement of system specifications (i.e., here it denotes the weakening of the preconditions and of the operation indeterminism towards the implementation level), there is always the possibility of keeping a B specification up to date. The B refinement allows the decomposition of system specification in layers and particular B clauses (e.g., use of SEES and IMPORTS clauses). This allows a modular system proof. New modules can be added to a B specification and existent ones may be assumed as being already proved; there is no need to totally reprove (i.e., recompute) the new specification. We therefore reuse an already proved specification. This aspect is very important as the security functionalities may be changed in a given system. Moreover, since specifying the system is an important step in our proposal, the link with the security policy must be established in a specific way. In this sense, the OrBAC philosophy considers that the security policy must be detached by functionality and by technology details at network level; and that changing the system architecture (hereinafter system or network with the same meaning) has no impact in policy definition. The same OrBAC policy may be specified for two different systems. Finally, the specification language must ensure the previous constraint: changing the security policy or the architecture should not trigger a new call for a total system proof.

We consider that the B-Method is suitable to achieve these purposes: in our approach the policy SEES the system it was designed for. The SEES clause in the B-Method makes the assumption that the *seen* system is already proved (i.e., the INVARIANTS of the SEEING module are being proved with the assumption that the INVARIANTS of the SEEN module are already proved). Let us notice that in our approach, we use the terminology *security property* as a synonym of *correct* deployment of a security policy in a system, meaning that it is achievable whenever certain specific security properties are verified. For instance, all network traffic between two network zones is protected if all traffic passes through an IPsec tunnel with certain parameters. If an IPsec tunnel is established and there is no IPsec tunnel anomaly [12] related to the current tunnel, the *integrity* security property is consequently verified. Hence, in a B specification there is the possibility of capturing such details at the INVARIANT clause level.

Some security requirements are dynamic or contextual. It is sometimes necessary to add new security functionalities to the given system. For example, some firewalls are upgraded with new functionalities (e.g., temporal functionality). Taking into account all security functionalities is out of our scope. We, therefore, address only some basic functionalities such as packet filtering, IPsec tunneling and signature-based Intrusion Detection. If further functionalities are added to the specification, their semantics must be reflected in particular SETS, CONSTANTS and consequently PROPERTIES clauses; but the main deployment algorithms should remain unaffected.

3.2   Policy and System Modeling

The starting point in deploying a security policy is a set of OrBAC abstract rules. A first assumption is that the abstract policy is consistent: no OrBAC conflicting rules. This is ensured by some pertaining tools like MotOrBAC which implements the conflict resolution described in [9].

The *context* definition may be related to a specific subject, action and object. Consequently, it is necessary to instantiate the corresponding subjects, actions and objects for each such contexts before deploying the OrBAC rules over each PEP. The abstract OrBAC rules, Permission(org, r, a, ν, c), must be brought to a concrete OrBAC expression [1], Is_permitted(org, s, α, o, c). Even if a large set of concrete security rules will have to be deployed, this may be the only option if the security requirements imply only such context definitions. At this point, we refer to the works in [22] that addressed the refinement problem: the OrBAC abstract expression towards a concrete one and using the B-Method (cf. Section 2). The works in [22] stopped at our stage, i.e., the link with the system. Therefore we will make a second assumption: the OrBAC concrete rules (i.e., Is_permitted(org, s, α, o, c)) are already available and they represent the input to our deployment process. The main entities to implement our approach are described as follows. The "security policy" is defined as the set of rules over the domain (Subjects $\times$ Actions $\times$ Objects $\times$ Contexts). "Subjects" and "objects" represent active and respectively passive entities in the network. A host in a subnetwork may be modeled as a subject in contrast to a web-server which may be seen as an object; not only the hosts/network components but also the clients and servers applications may be seen as subject-object entities. The "actions" are defined as network services (e.g., http and https are actions of the same abstract activity, *web*). We should also include the "contexts" in which some rules are activated. These may be bound in *hard* with some functionalities. For example, the *protected* context relies on IPsec functionalities and the *warning* context on IDS functionality. *Is_permitted(s, α, o, default)* is activated in the *default* context only if a path from s to o exists, so the firewalls on this path have to open some ports corresponding to the action α. Finally, the approach also includes all those interesting "nodes" in the network (subjects and objects) will appear as nodes in a connected graph. Those having security functionalities are the PEPs. A PEP may also be a subject or an object.

Figure 1 depicts a sample network in which an access control policy must be deployed. The system is modeled as a graph (cf. Figure 2). In real networks, each link may have a real cost or weight (e.g., an overhead required to send IP packets over the link and inversely proportional to the bandwidth of the link). A routing protocol like the OSPF, *Open Shortest Path First* [rfc 2328], establishes routes in choosing, for example, the shortest paths (i.e., the less expensive). We choose positive integers for the link costs and we assume that IP datagrams always follow the shortest path between two points. In such a system, the PEPs must be enforced with the right decisions corresponding to each security rule. For didactical reasons we take into account only permissions. One may choose to enforce the same rule in all PEPs having the same functionalities: this is what we call a redundancy anomaly and this is what we try to avoid. Our configuration approach is the following: for each permission, only the *interesting* PEPs must be identified and enforced. This leads us to consider an algorithm for selecting

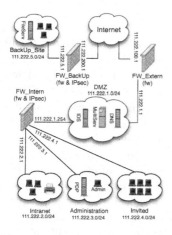
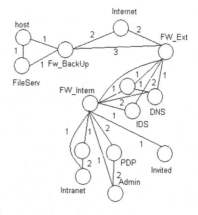

Fig. 1. The Real Architecture   **Fig. 2.** The Corresponding Graph

the interesting PEPs: the well placed PEPs for each security rule. For example, a rule in a default context takes into account only the firewalls on a certain path. In a protected context, two PEPs must configure an IPsec tunnel and all firewalls on the tunnel path must permit the establishment of the tunnel. Regarding the warning context, the most down-stream IDS (i.e., the closest IDS to the destination) is enough and more efficient to spoofing attacks than the most up-stream IDS (i.e., the closest to the source).

We aim to formally implement this approach in B and go as far as possible towards the IMPLEMENTATION level. In this manner we will capture all interesting details for the security property enouncement. In the following sections we consider the policy at concrete OrBAC level as described above. We deal with a system where some distributed nodes have security functionalities (PEPs) and some are either active entities (subjects) or resources (objects) or both.

4   Policy Deployment Process: Formal Specification

Not all B machines will be carried out to an IMPLEMENTATION level: the *Policy* and *Network* (cf. Figure 4) machines should be instantiated for each scenario. The data these machines manipulate does not require highly specialized mathematical objects: only lists/sequences must be provided. But the other machines we present will have an IMPLEMENTATION structure. We describe in detail the Policy and Network (system) machines; they will be incorporated to our model development as a result of an IMPORTING machine: *Deployment* machine which also imports other machines necessary to our process. The *Path* machine will implement a tracing path algorithm necessary to select the well placed PEPs for each security rule which are then updated by the *UpdatePep* machine (cf. Figure 3).

4.1 Policy and Network Machines

The SUBJECTS, ACTIONS and OBJECTS will represent deferred sets but we prefer for the moment a concrete/enumerated set of CONTEXTS: *default, protected, user-defined* and *other*. The set of permissions to be deployed as well as the matching *nodes-subjects, nodes-objects* may be considered as CONSTANTS. They will be defined via some relations, more precisely functions. As already mentioned we prefer to bind in *hard* a context to a security functionality; we model this by the *matching* relation. Moreover, the user will be given the possibility of defining other types of context activation. For instance, a certain user-defined context may impose a hub-and-spoke tunnel configuration so the user must be able to manually indicate the hub and the spokes (nodes in the network). We model this by the *context* constant relation. Besides, the permissions are progressively read in the deployment process. An abstract variable is consequently necessary, the Read_Permissions. All these semantics will be summed up at PROPERTIES level.

Regarding the dynamic part of the Policy machine: the Read_Permissions is initialized with the empty set and some simple operations are necessary (1) to read and return a permission (read_permission) and (2) to read the attributes (subject, object, action and context) of a permission (read_data_in_permission). The INVARIANT is a simple one, it checks the variable type. Other inquiry operations (no_read_permission, no_more_permissions) simply return *true* or *false*.

Concerning the Network machine, we can envision the following two options. We can use an abstract machine encapsulating a node, say the Node machine. It should contain at least the node functionalities as a deferred SET. Node could be imported in our project by renaming: the project will therefore contain as many renamed Node machines as the existent ones in the real network. In the same project, a different SEEN machine will define, via a constant, the network

The Project Organization

A Dependency Graph is automatically obtained from a B project in Atelier B if no machine sharing rules (i.e., via SEES, IMPORTS clauses) are violated.

We deliberately do not charge the graph with the other machines necessary to the final implementation of Weighted_Forest and Priority_Queue. Their implementation is similar to the one in [12] and they do not reveal any important security details.

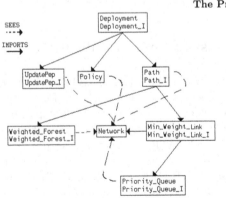

Fig. 3. Dependency Graph

topology (e.g., the nodes connections). The second option is to define a machine Network containing the topology description from the beginning. This way, we do not require other machines to carry out the network topology. Network will simply be imported only once in the project.

We choose the second solution. The Network machine models a graph: a non-empty set of *Nodes*, a set of *Links* ∈ *Nodes* × *Nodes* and a *weight* function binding a link to a natural number. We choose to identify each node by a natural number in the sequence 1..*nn*, *where* nn = card(*Nodes*). These are constants that require a refinement (i.e., valuation) at the IMPLEMENTATION level with concrete data for each different topology. The graph is undirected, so *Links* ∩ $Links^{-1}$ = ∅. The connectedness assumption is caught by All_Links*= (*Nodes*) × (*Nodes*) (where * stands for the transitive closure of relations). Each node may have security functionalities: *functionality* ∈ *Nodes* ↔ *FUNCTIONALITIES* (a constant relation). Some definitions are also necessary when a path is traced in the network: all links having a common node and the cost/weight of a set of links. These are the λ-functions *cost* and *neigh_nodes*.

When choosing the variables of the Network machine we take into account not only the network parsing aspect but also the nodes involved in the construction of paths (shortest paths) given a source node and a destination one. We, therefore, introduce some *processed* nodes (*PPnodes*) and links variables necessary in our shortest path algorithm. The INVARIANT of the Network machine acts on the variables type. We follow a generous style in specifying the operations: almost each operation has some preconditions. The generous style, in contrast to the *defensive* style which considers some internal operation tests (e.g., IF, SELECT substitutions), is more in the spirit of a B specification ([2]) as it demands prior design and specification. Such operations may be called from somewhere else (i.e., operations of other machines) and their preconditions must be verified; otherwise, they may not *terminate*.

Policy SEES Network (cf. Figure 4). Although the constants of Network may appear in the operations of a SEEING machine (e.g., Policy) its variables may be read-only. No operation of a SEEN machine can be called from within a SEEING machine. Therefore, Network is not aware of the fact that it is seen. Even if there are specific proof-obligations generated as a result of a SEES clause ([2]), the invariants of the SEEN machine are considered already proved. If we change the Policy for the same Network, the latter is once and independently proved. A SEEN machine, and consequently Network, may be imported only once somewhere in the project. We pay attention and we import Network only in a machine that really necessitates more than read-only variable references. This is illustrated in Figure 3.

4.2   A Tracing Path Algorithm

The role of a tracing path algorithm in our development is to find the security devices that must enforce each policy rule. These security devices must have the right functionalities and must be well placed in the network. If the right functionalities are indirectly designated by the OrBAC contexts (i.e., *default* - firewall, *protected* - IPsec and firewalls, *warning* - IDS), finding the well placed device in the network is not obvious. Nevertheless things are getting simpler if we consider that IP packets follow the shortest path in the network. One could say this is a severe assumption, but conciliating a given routing policy with our deployment process is a simpler matter: it suffices to take into account the few hops a route may involve.

Therefore, we use Dijkstra's shortest path algorithm. Implementing such an algorithm in imperative languages is not too difficult but it is not obvious using the B-Method. There are already B algorithms for deriving spanning trees, [11], [3], but none for shortest-path trees. We base our path derivation algorithm on the works in [11]; we could not totally reuse their method: the shortest path may not go along the minimum spanning tree. Moreover, shortest-path tree changes as a result of choosing different source and destination nodes. Some implementation details in [11] concerning the priority-queues turned out to be extremely useful: we reused them although the lists would have been much easier to manipulate. However, we mention that the project in [11] violates a sharing rule: a SEEN machine must be imported once but IMPORTS must not introduce loops in the project. We believe their error is due to their prover which did not check on machine sharing rules.

Path Machine, Weighted Forest, Min Weight Link: Path machine SEES Network whose constants (Nodes, Links and weight) are used. We need a definition of a path in the network. But as a pre-requisite we have to formalize the notion of a tree, more precisely of a spanning-tree: a forest with (n - 1) links, where n = card(Nodes) and a forest is a cycle-free set of links. We also need a definition of the set of paths from a source node to a destination node: all (i.e., the union of) adjacent links with the source and destination as extremity nodes. We are therefore interested in selecting the less-expensive path in this set of paths. This will represent the shortest path which will be simply selected once the shortest-path tree is generated (shortest_path_tree). In what follows we introduce only some specific B details we faced when designing Path machine and its implementation (the termination of the shortest_path_tree operation is ensured by our assumption of a connected graph).

Implementing the Path machine with the previous specifications would be difficult. The IMPLEMENTATION will therefore import two machines: *Min_Weight_Link* machine, to find the minimum link weight in a set of links adjacent to some processed nodes (Dijsktra's algorithm) and *Weighted_Forest* machine, to build the tree as a union of links. The tree is noted LL which is an abstract variable of Weighted_Forest. LL finally represents the shortest-path-tree. Due to space limitations we do not go further with our algorithm. The complete implementation of Weighted_Forest and Min_Weight_Link has several hundreds of B code lines. We mention only that we used the Abstract_Constants clause in order to avoid the error in [11] regarding the SEES and IMPORTS clauses: if an abstraction SEES a machine, all the further refinements must also

MACHINE *Policy*
SEES *Network*
SETS
    SUBJECTS;
    ACTIONS;
    OBJECTS;
    CONTEXTS={*default,   prot,   user_def, other_ctx*}
DEFINITIONS
    Nodes==1 .. *nn*
CONSTANTS
    context, Permissions, Subject, Object, matching
PROPERTIES
    context $\in$ *CONTEXTS* $\leftrightarrow$ *seq*(*Nodes*) $\wedge$
    Permissions $\in$ $\mathcal{P}$(*SUBJECTS* $\times$
ACTIONS $\times$ *OBJECTS* $\times$ *CONTEXTS*) $\wedge$
    matching $\in$ *CONTEXTS* $\leftrightarrow$ *FUNCTIONAL-ITIES* $\wedge$ *Subject* $\in$ *SUBJECTS* $\rightarrow$ *Nodes* $\wedge$
    Object $\in$ *OBJECTS* $\rightarrow$ *Nodes*

VARIABLES
    Read_Permissions
INVARIANT
    Read_Permissions $\subseteq$ *Permissions*
INITIALISATION
    Read_Permissions:= $\emptyset$

OPERATIONS
    no_read_permission =
      Read_Permissions:= $\emptyset$;

    permission $\leftarrow$ **read_permission** =
      PRE *Read_Permissions* $\neq$ *Permissions*
      THEN ANY *per* **WHERE**
per $\in$ *Permissions*-*Read_Permissions* **THEN**
      permission, Read_Permissions :=
per, Read_Permissions $\cup$ {*per*}
      END   **END**;

    bb $\leftarrow$ **no_more_permissions** =
bb:=**bool**(*Read_Permissions*=*Permissions*);

ss,aa,oo,cc $\leftarrow$ **read_data_in_permission**(*pp*) =
PRE *Read_Permissions* $\neq$ $\emptyset$ $\wedge$
    pp $\in$ *Read_Permissions* **THEN**
ANY *sub, act, obj, ctx* **WHERE** *sub* $\in$ *SUB-JECTS* $\wedge$ *act* $\in$ *ACTIONS* $\wedge$ *obj* $\in$ *OBJECTS* $\wedge$ *ctx* $\in$ *CONTEXTS* $\wedge$ {*sub* $\mapsto$ *act* $\mapsto$ *obj* $\mapsto$ *ctx*}={*pp*} **THEN**
        ss, aa, oo, cc:=*sub, act, obj, ctx*
      END
    END

END /*Policy.mch*/

MACHINE *Network*
SETS
    FUNCTIONALITIES = {*fw, ipsec, ids, other*}
CONSTANTS
    nn, Links, weight, functionality
DEFINITIONS
    Nodes==1 .. *nn*;
    All_Links==*Links* $\cup$ *Links* $^{-1}$;
    cost== λ *LL*.(*LL* $\in$ $\mathcal{P}$ (*Links*) | $\sum$ *link*.(*link* $\in$ *LL* | *weight*(*link*)));
    neigh_nodes== λ *LL*.(*LL* $\in$ *Nodes* | ($\bigcup$ *ii*.(*ii* $\in$ *Nodes* $\wedge$ *LL* $\mapsto$ *ii* $\in$ *Links* | {*LL* $\mapsto$ *ii*})) $\cup$ ($\bigcup$ *jj*.(*jj* $\in$ *Nodes* $\wedge$ *jj* $\mapsto$ *LL* $\in$ *Links* | {*jj* $\mapsto$ *LL*})))
PROPERTIES
    nn $\in$ **NAT1** $\wedge$ *Links* $\in$ *Nodes* $\leftrightarrow$ *Nodes* $\wedge$
weight $\in$ *Links* $\rightarrow$ **NAT** $\wedge$ *Links* $\cap$ *Links* $^{-1}$ = $\emptyset$
$\wedge$ *All_Links**=(*Nodes*) $\times$ (*Nodes*) $\wedge$
    card(*Links*) $\in$ **NAT1** $\wedge$ *functionality* $\in$ *Nodes* $\leftrightarrow$ *FUNCTIONALITIES* $\wedge$ (*cost*)(*Links*) $\in$ **NAT**

VARIABLES
    Read_Links, Read_Nodes, Neighbors, PPnodes, Read_Neighbors
INVARIANT
    Read_Links $\in$ $\mathcal{P}$ (*Links*) $\wedge$ *Read_Nodes* $\in$ $\mathcal{P}$ (*Nodes*) $\wedge$ *Neighbors* $\in$ $\mathcal{P}$ (*Links*) $\wedge$ *PPnodes* $\in$ $\mathcal{P}$ (*Nodes*) $\wedge$ *Read_Neighbors* $\in$ $\mathcal{P}$ (*Links*)
INITIALISATION
    Read_Links, Read_Nodes, Neighbors, PPnodess, Read_Neighbors:= $\emptyset$, $\emptyset$, $\emptyset$, $\emptyset$, $\emptyset$

OPERATIONS /* for space limitation reasons, not all operations are detailed */
    node, func $\leftarrow$ **read_node** =
      PRE *Read_Nodes* $\neq$ *Nodes* **THEN**
      ANY *nod* **WHERE** *nod* $\in$ *Nodes* $\wedge$ *nod* $\notin$ *Read_Nodes* **THEN**
      node, func, Read_Nodes:=*nod, functionality*[{*nod*}], *Read_Nodes* $\cup$ {*nod*}
      END   **END**;

    uu, vv, ww $\leftarrow$ **read_link** =
      PRE *Read_Links* $\neq$ *Links* **THEN**
      ANY *ii, jj* **WHERE** *ii* $\in$ *Nodes* $\wedge$ *jj* $\in$ *Nodes* $\wedge$ (*ii,jj*) $\in$ *Links*-*Read_Links* **THEN**
      uu, vv, ww, Read_Links:=*ii, jj, weight*(*ii, jj*), *Read_Links* $\cup$ {*ii* $\mapsto$ *jj*}
      END   **END**;

    new_neighbors(*uu*) = **PRE** *uu* $\in$ *Nodes* $\wedge$ (*Neighbors* $\cap$ (*neigh_nodes*)(*uu*)) $\neq$ $\emptyset$ $\wedge$ *uu* $\notin$ *PPnodes* **THEN** ... /*set new Neighbor Links*/

... /*other operations*/

END /*Network.mch*/

Fig. 4. Policy and Network Machines

SEES this machine and the final IMPLEMENTATION cannot IMPORTS the seen machine.

4.3 Deployment Implementation and Security Properties

The root machine of our model is the Deployment machine (cf. Figure 5). Its abstract specification is quite simple: there is only a Boolean concrete variable, *deployment_ok* modified by an operation, *deploy*. The refinement of this operation is based on other operations of the IMPORTED machines Policy, Path and UpdatePep. Network is imported in our model indirectly, via the Path machine. By using the IMPORTS clause, allowed only from within an IMPLEMENTATION there are specific obligation-proofs generated for the IMPORTING machine. We deliberately leave the IF substitution unfinished: there are tests concerning the existence of a path in the network according to the type of context. Therefore, *exists_path*, a boolean variable of the Path machine, is valued in function of several other variables: the security functionalities of the source and destination nodes (e.g., the IPv6 protocol incorporates the IPsec suite but this functionality may not be considered in the IPsec tunnel extremities), the security functionalities in their neighborhood (e.g., for IDS rules) or the security functionalities of the whole path (*path_set*) between the source and the destination. These tests are simple and rely on the definition and implementation of the Path machine. Finally, the *UpdatePep* machine stores, for each PEP, the security configuration as a set of rules $\{sub \mapsto act \mapsto obj \mapsto ctx\}$ modified via the *update_pep* operation. The concrete *deployment_ok* variable respects the data types required in an IMPLEMENTATION. It needed no further refinement and we defined it as a CONCRETE_VARIABLE in the abstraction.

4.4 Security Properties

A security property is generally expressed at a more abstract level than the security requirements. A security property may rely on the correct enforcement of several security requirements. Moreover, a property may still not be verified after the deployment of all security rules. Often, a property violation is the result of anomalies in deploying the policy. Our refinement approach is a property-aware one: the target properties determine the enforcement of the security devices. In the implementation of the Deployment machine (cf. Figure 5), we denote by P_1-P_8 some of the most interesting application-independent properties the policy deployment process should verify. We do not claim to achieve a thorough analysis of security properties: some may be enounced at higher levels ([6]) and some may be identified from specific security requirements ([15]).

- **Completeness:** Captured by INVARIANT P_1, this property states that if the network path from a subject to an object is correctly computed (i.e., it exists and the security devices belonging to this path have the right functionalities with respect to the context) the security rule may and will be deployed.

IMPLEMENTATION *Deployment_I*
REFINES *Deployment*
IMPORTS *Path, Policy, UpdatePep*
INVARIANT

P_1 $\Big\{ (deployment\_ok = \text{true}) \Leftrightarrow (exists\_path=\text{true}) \land$

P_2 $\begin{cases} \forall(sub,\ act,\ obj).(sub \in SUBJECTS \land act \in ACTIONS \land obj \in OBJECTS \land \\ (sub \mapsto act \mapsto obj \mapsto prot) \in PERMISSIONS \Rightarrow \neg\exists(n1,\ n2).((n1 \mapsto n2) \in Links \land \\ \{n1, n2\} \subseteq shortest\_path(Subject(sub),\ Object(obj\ )) \land \\ (sub \mapsto act \mapsto obj \mapsto default) \in config[\{n1\}] \cap config[\{n2\}])) \land \end{cases}$

P_3 $\begin{cases} \forall(sub,\ act,\ obj).(sub \in SUBJECTS \land act \in ACTIONS \land obj \in OBJECTS \land \\ (sub \mapsto act \mapsto obj \mapsto default) \in PERMISSIONS \Rightarrow \\ \forall(node).(node \in Nodes \land fw \in functionality\ [\{node\}] \land \\ node \in shortest\_path(Subject(sub),\ Object(obj\ )) \Rightarrow \\ (sub \mapsto act \mapsto obj \mapsto default) \in config[\{node\}])) \land \end{cases}$

P_4 $\begin{cases} \forall(sub,\ act,\ obj).(sub \in SUBJECTS \land act \in ACTIONS \land obj \in OBJECTS \land \\ (sub \mapsto act \mapsto obj \mapsto default) \in PROHIBITIONS \Rightarrow \\ \forall(node).(node \in Nodes \land fw \in functionality\ [\{node\}] \land \\ node \in path(Subject(sub),\ Object(obj\ )) \Rightarrow \\ (sub \mapsto act \mapsto obj \mapsto default) \in config[\{node\}])) \land \end{cases}$

P_5 $\begin{cases} \forall(sub,\ act,\ obj).(sub \in SUBJECTS \land act \in ACTIONS \land obj \in OBJECTS \land \\ (sub \mapsto act \mapsto obj \mapsto authent) \in PERMISSIONS \Rightarrow (authentication \in history[\{sub\}])) \end{cases}$

INITIALISATION
    deployment_ok:=true
OPERATIONS
    deploy = **VAR** *permis, bb, kp* **IN no_read_permission;**
        bb ← **no_more_permissions;** *kp:*=0;
    WHILE *bb*=false **DO**
        permis ← **read_permission;** *kp:*=*kp*+1;
    VAR *suj, act, obj, cont, src, dest, path_set* **IN**
suj, act, obj, cont ← *read_data_in_permission(permis);*
        src:=*Subject(suj); dest:*=*Object(obj);*
        IF *dest* ≠ *src* **THEN**
            shortest_path_tree*(src);* *path_set* ←**shortest_path***(src, dest);*
            ...
            update_pep*(src, dest, path_set);* *deployment_ok:*=true
        ELSE   *bb:*=**true** /*loop exit*/   **END   END**
        INVARIANT   *bb* ∈ **BOOL** ∧

P_6 $\Big\{ (cont = default) \Rightarrow (exists\_path = true) \land$

P_7 $\begin{cases} (cont = default \lor cont = logging\ ) \Rightarrow (exists\_path=true \land \\ \exists\ node.(node \in Nodes \land node \in shortest\_path(src, dest) \land \\ fw \in functionality[\{node\}])) \land \end{cases}$

P_8 $\begin{cases} (cont = prot) \Rightarrow (exists\_path = true\ ) \land \\ ipsec \in (functionality[\{src\}] \cap functionality[\{dest\}]) \end{cases}$

    VARIANT card*(Permissions)-kp*   **END   END** /*deploy*/
END /*Deployment_I*/

Fig. 5. Deployment Implementation

- **Accessibility (and Inaccessibility):** Property P_6 states that for each permission rule, a subject is able to access an object with respect to the policy. Thus, there must be a path between the subject and the object network entities and if this path involves some firewalls, they must all permit the action the subject is supposed to realize on the object. In this manner, the *default* context is activated (this may be seen as a *minimal* context). However, P_6 must be seen as a *partial* accessibility property: it is verified at each WHILE loop iteration, i.e., it does not take into account all deployed rules. In order to guarantee the *global* accessibility, P_3 relies on the correct deployment of all permissions. We simplified the notation: given that the operation *shortest_path(src, dest)* returns a set of links called *path_set*, we should have written: *node* $\in$ *path_set*. In P_3 we use *config* which is defined in the UpdatePep machine: *config* $\in$ *Nodes* $\leftrightarrow$ *PERMISSIONS* and *config[{ni}]*, the image of the $\{ni\}$ set under *config*, is the set of rules already deployed over the node *ni*.

  Following the same reasoning we can also enounce the global inaccessibility property, P_4: there should be no open *path* from the subject to the object. The *path* variable is given in the Path machine and regroups a set of links from a source node to a destination one.
- **All traffics are regulated by firewalls:** Property P_7 is also interesting when there is a new context called *logging*: this context is managed by those devices with a logging functionality as today's most popular firewalls.
- **Integrity and confidentiality property:** This property is related to the establishment of IPsec tunnels. It ensures the extremities of the IPsec tunnel. Moreover, particular IPsec configurations may include recursive encapsulation of traffic on a path. Verifying this property begins at higher levels: if no OrBAC security rule is enounced with a protected (*prot*) context, no further verification is necessary. To ensure the protected context activation, a configuration of an IPsec tunnel is necessary. If no specific information concerning the IPsec tunnel establishment is provided, we may suppose the following two cases: (1) the subject/source and the object/destination are IPsec enabled (e.g., IPv6 nodes and end-to-end tunnel) or (2) at least one node in their neighborhood (e.g., site-to-site tunnel) has IPsec functionalities. For the first case, it suffices to check on the IPsec functionalities on both the subject and the object nodes and this is captured by the P_8 property of the WHILE loop. The second case is handled as follows: in one of the IMPORTED machines on the Weighted_Forest development branch, we provide an operation *predec(node)* which returns the *precedent* node in the current shortest-path from the source (src) node to the destination (dest) node. Via PROMOTE clauses, the operation may be called by *higher* IMPORTING machines, including the Deployment machine. We consequently check on the predec(dest) and predec^{-1}(src) nodes as in the P_8 formula.

  There is a further case that cannot be addressed at the WHILE loop INVARIANT level (i.e., after each iteration) because it relies on the correct enforcement of all IPsec tunnels in the network. Figure 6 shows an example of successive encapsulations on a site-to-site topology. It may result in violating the confidentiality property. It also shows an example where the source-destination traffic is twice encapsulated: by n1 (IPsec tunnel mode between n1 and n2) and by n3 (between n3 and n4). The configurations of n2 and n3 neighbor nodes include a security rule in a default context (i.e., $\{sub \mapsto act \mapsto obj \mapsto default\}$) allowing the IP traffic to pass the section n2-n3 with no encapsulation: the confidentiality is not preserved in this topology (i.e., between n1 and n4). Such an anomaly is the result of deploying separately the IPsec tunnels and consequently it cannot be controlled by the WHILE INVARIANT after each iteration. Nevertheless, the main INVARIANT of the Deployment machine is not reproved during these iterations but after the loop termination and consequently the IPsec anomaly can be dealt with only at this level. The P_2 formula, in logical conjunction with the *completeness* property, accomplishes the integrity and confidentiality property.

Fig. 6. Chained IPsec Tunnels

- **Authentication:** P_5 is interesting if we deal with an *authentication* context: an action that cannot be applied in a certain context unless an authentication process is achieved. We can verify these cases by providing a variable that records the actions realized by the subject concerned with the authentication. We, therefore, impose a workflow constraint (*history*[$\{sub\}$] is the set of actions that *sub* realized, with *history* $\in SUBJECTS \leftrightarrow ACTIONS$).

5   Discussion

The B project depicted in this paper was realized using Atelier B v4.0. Validating a B project consists in proving the Proof Obligations (POs) automatically generated after analyzing and type-checking the entire project. The functional correctness of each machine is validated separately with respect to the specific B inter-machine clauses (SEES, IMPORTS, etc.). The current project was validated with the assumption of a conflict-free OrBAC policy and of a correct system architecture: no lack of security functionalities in the security components placed on the shortest-paths. This leads us to conclude that the outgoing algorithms are correct with respect to the security properties we considered. The number of POs automatically generated for each machine varied based on the operational complexity: from 2 for the Policy machine which involved very simple operations to 272 for the Min_Weight_Link implementation; for the latter one, 110 POs were automatically discharged, the rest being interactively proved.

The choice of the OrBAC model and of the B-Method was motivated by the type of applications that we address in this paper: the deployment of access control security policies. However, our work shows some limitations. On the one hand, our approach focuses on the deployment of policies in systems with an already existing set of security devices. The appropriate deployment of the access control policy is closely related to the interconnections and the capabilities of these security devices. As we avoid the intra- and inter-component anomalies ([13]), there may be unaccomplished security requirements because of a deficient security device capability. Thus, an improvement to our approach would be to find, for a given system and a given security policy, the best security architecture so that all security requirements be met. On the other hand, the type of security requirements may also induce a limitation to our approach. As long as we consider only access control requirements, the B-Method is very efficient. However, temporal logic specifications cannot be addressed with the B Method. Therefore, except for the specific case of authentication, the security requirements involving a trace-like modeling (an ordered set of actions to be realized by a subject on an object) cannot be addressed with the B Method. The authentication can be dealt with since the subject needs to accomplish a single (previous) action (modeled by a *provisional — history — OrBAC context*) before gaining the access.

6   Conclusions

The configuration of security devices is a complex task. A wrong configuration leads to weak security policies, easy to bypass by unauthorized entities. The existence of reliable automatic tools can assist security officers to manage such a cumbersome task. In this paper, we established a formal frame for developing such tools. Our proposal allows the administrator to formally specify security requirements by using an expressive access control model based on OrBAC [1]. A tool which is proved using the B-Method may therefore implement the so-called *downward* process: the set of algorithms realizing the translation of an OrBAC set of rules into specific devices configurations. Not only the job of administrators is simplified, but they know for certain what security properties are verified at the end.

Acknowledgments. This work has been supported by a grant from the Brittany region of France and by the following projects: POLUX ANR-06-SETIN-012, SEC6 Project, TSI2007-65406-C03-03 E-AEGIS, and CONSOLIDER CSD2007-00004 "ARES".

References

1. Abou el Kalam, A., Baida, R.E., Balbiani, P., Benferhat, S., Cuppens, F., Deswarte, Y., Miège, A., Saurel, C., Trouessin, G.: Organization Based Access Control. In: IEEE 4th Intl. Workshop on Policies for Distributed Systems and Networks, Lake Come, Italy, pp. 120–131 (2003)
2. Abrial, J.R.: The B-Book — Assigning Programs to Meanings. Cambridge University Press, Cambridge (1996)
3. Abrial, J.R., Cansell, D., Méry, D.: Formal Derivation of Spanning Trees Algorithms. In: Bert, D., Bowen, J.P., King, S. (eds.) ZB 2003. LNCS, vol. 2651, pp. 457–476. Springer, Heidelberg (2003)

4. Autrel, F., Cuppens, F., Cuppens-Boulahia, N., Coma, C.: MotOrBAC 2: A security policy tool. In: SAR-SSI 2008, Loctudy, France (2008)
5. Bartal, Y., Mayer, A., Nissim, K., Wool, A.: Firmato: A novel firewall management toolkit. In: IEEE Symposium on Security and Privacy, pp. 17–31 (1999)
6. Benaissa, N., Cansell, D., Méry, D.: Integration of Security Policy into System Modeling. In: Julliand, J., Kouchnarenko, O. (eds.) B 2007. LNCS, vol. 4355, pp. 232–247. Springer, Heidelberg (2006)
7. Capretta, V., Stepien, B., Felty, A., Matwin, S.: Formal correctness of conflict detection for firewalls. In: ACM workshop on Formal methods in security engineering, FMSE 2007, Virginia, USA, pp. 22–30 (2007)
8. Casassa Mont, M., Baldwin, A., Goh, C.: POWER prototype: towards integrated policy-based management. In: Network Operations and Management Symposium, USA, pp. 789–802 (2000)
9. Cuppens, F., Cuppens-Boulahia, N., Ben Ghorbel, M.: High level conflict management strategies in advanced access control models. Electronic Notes in Theoretical Computer Science (ENTCS) 186, 3–26 (2007)
10. Damianou, N., Dulay, N., Lupu, E., Sloman, M.: The ponder policy specification language. In: Sloman, M., Lobo, J., Lupu, E.C. (eds.) POLICY 2001. LNCS, vol. 1995, pp. 18–38. Springer, Heidelberg (2001)
11. Fraer, R.: Minimum Spanning Tree. In: FACIT 1999, pp. 79–114. Springer, Heidelberg (1999)
12. Fu, Z., Wu, S.F., Huang, H., Loh, K., Gong, F., Baldine, I., Xu, C.: IPSec/VPN Security Policy: Correctness, Conflict Detection and Resolution. In: Policy 2001 Workshop, Bristol, UK, pp. 39–56 (2001)
13. Garcia-Alfaro, J., Cuppens, N., Cuppens, F.: Complete Analysis of Configuration Rules to Guarantee Reliable Network Security Policies. International Journal of Information Security 7(2), 103–122 (2008)
14. Jürjens, J.: Secure Systems Development with UML. Springer, New York (2004)
15. Jürjens, J., Schreck, J., Bartmann, P.: Model-based security analysis for mobile communications. In: 30th international conference on Software engineering, Leipzig, Germany, pp. 683–692 (2008)
16. Laborde, R., Kamel, M., Barrere, F., Benzekri, A.: Implementation of a Formal Security Policy Refinement Process in WBEM Architecture. Journal of Network and Systems Management 15(2) (2007)
17. Ioannidis, S., Bellovin, S.M., Ioannidis, J., Keromitis, A.D., Anagnostakis, K., Smith, J.M.: Virtual Private Services: Coordinated Policy Enforcement for Distributed Applications. International Journal of Network Security 4(1), 69–80 (2007)
18. Ponsini, O., Fédèle, C., Kounalis, E.: Rewriting of imperative programs into logical equations. In: Sci. Comput. Program., vol. 54, pp. 363–401. Elsevier North-Holland, Inc., Amsterdam (2005)
19. Preda, S., Cuppens, F., Cuppens-Boulahia, N., Alfaro, J.G., Toutain, L., Elrakaiby, Y.: A Semantic Context Aware Security Policy Deployment. In: ACM Symposium on Information, Computer and Communication Security (ASIACCS 2009), Sydney, Australia (March 2009)
20. Sandhu, R., Coyne, E.J., Feinstein, H.L., Youman, C.E.: Role-Based Access Control Models. IEEE Computer 29(2), 38–47 (1996)
21. Unal, D., Ufuk Çaglayan, M.: Theorem proving for modeling and conflict checking of authorization policies. In: Proceedings of the International Symposium on Computer Networks, ISCN, Istanbul, Turkey (2006)
22. ACI DESIRS project: DÉveloppement de Systèmes Informatiques par Raffinement des contraintes Sécuritaires

Category-Based Authorisation Models: Operational Semantics and Expressive Power

Clara Bertolissi[1] and Maribel Fernández[2]

[1] LIF, Université de Provence, Marseille, France
`Clara.Bertolissi@kcl.ac.uk`
[2] King's College London, Dept. of Computer Science, London WC2R 2LS, U.K.
`Maribel.Fernandez@kcl.ac.uk`

Abstract. In this paper we give an operational specification of a meta-model of access control using term rewriting. To demonstrate the expressiveness of the meta-model, we show how several traditional access control models, and also some novel models, can be defined as special cases. The operational specification that we give permits declarative representation of access control requirements, is suitable for fast prototyping of access control checking, and facilitates the process of proving properties of access control policies.

Keywords: Security Policies, Access Control, Operational Semantics, Term Rewriting.

1 Introduction

Over the last few years, a wide range of access control models and languages for access control policy specification have been developed, often motivated by particular applications. In contrast, Barker [4] proposes a general meta-model for access control based on a small number of primitive notions, which can be specialised for domain-specific applications. This approach has advantages: for example, by identifying a core set of principles of access control, one can abstract away many of the complexities that are found in specific access control models; this, in turn, helps to simplify the task of policy writing.

In this paper, we provide a formal specification of Barker's meta-model $\mathcal{M}$, using a general formalism, based on term rewriting, that enables access control policies to be defined in a declarative way and permits properties of access control policies to be proven. Term rewriting systems present many advantages as a specification tool: they have a well-studied theory, with a wealth of results that can be applied to the analysis of policies, and several rewrite-based programming languages are available (see, e.g., Maude [10]). We use distributed term rewriting [9] to provide a formal semantics for policy specification, an operational semantics for access request evaluation, and a basis for a prototype

F. Massacci, D. Wallach, and N. Zannone (Eds.): ESSoS 2010, LNCS 5965, pp. 140–156, 2010.

implementation. Indeed, term rewrite systems can be seen as a formal, executable specification.

A key aspect of our approach, following [4], is to focus attention on the notion of a *category*.[1] A category (a term which can, loosely speaking, be interpreted as being synonymous with, for example, a sort, a class, a division, a domain) is any of several fundamental and distinct classes to which entities or concepts belong. We regard categories as a primitive concept and we view classic types of groupings used in access control, like a role, a security clearance, a discrete measure of trust, etc., as particular instances of the more general notion of category.

To demonstrate the expressive power of the category-based meta-model, we show how a range of access control models can be defined as specific instances of the meta-model. In particular, we show how policies defined in terms of the ANSI hierarchical role-based access control (H-RBAC) model [2], a mandatory access control (MAC) model [7], and the event-based access control (DEBAC) model [9], can be represented in our framework. We also describe how a variety of extended forms of these models may be specified, in terms of notions like time and location, and how a number of novel access control models can be derived as particular cases of the meta-model.

Summarising, the main contributions of this paper are:

- a declarative, rewrite-based specification of the category-based access control model $\mathcal{M}$, together with a formal operational semantics for access request evaluation, for access control policies specified using $\mathcal{M}$;
- a technique to prove totality and consistency of access control policies, by proving termination and confluence of the underlying term rewriting system;
- the encoding of well-known access control models in $\mathcal{M}$, to demonstrate the expressive power of the meta-model: specifically, we show that the H-RBAC, MAC and DEBAC models can be derived as instances of $\mathcal{M}$, as well as access control models with time and location constraints, the Chinese Wall policy model and access control models based on trust.
- the definition of novel access control models as instances of $\mathcal{M}$, such as a generalisation of DEBAC that we exemplify in a banking scenario.

The remainder of the paper is organised in the following way. In Section 2, we recall some basic notions in term rewriting, and we describe the main features of the access control meta-model. In Section 3, we specify the meta-model as a term rewriting system, give examples of general policy representation in terms of the meta-model, and describe techniques for proving properties of access control policies. In Section 4, we show how several different access control models and associated policies, e.g., RBAC policies and DEBAC policies, can be understood as special cases of the meta-model. In Section 5, we discuss related work. In Section 6, conclusions are drawn, and further work is suggested.

---

[1] From the Greek word "kategoria", which was used by Aristotle to distinguish the types of questions that may be asked of an entity with a view to categorising the entity.

2   Preliminaries

In order to make the paper reasonably self-contained, we recall some basic notions and notations for first-order term rewriting and the category-based authorisation meta-model that will be used in the rest of the paper. We refer the reader to [3] and [4] if additional information is required.

Term Rewriting. A *signature* $\mathcal{F}$ is a finite set of *function symbols*, each with a fixed arity. $\mathcal{X}$ denotes a denumerable set of *variables* $X_1, X_2, \ldots$. The set $T(\mathcal{F}, \mathcal{X})$ of *terms* built up from $\mathcal{F}$ and $\mathcal{X}$ can be identified with the set of finite trees where each node is labelled by a symbol in $\mathcal{F} \cup \mathcal{X}$ such that a node labelled by $f \in \mathcal{F}$ must have a number of subtrees equal to the arity of f, and variables are only at the leaves. *Positions* are strings of positive integers denoting a path from the root to a node in the tree. The *subterm* of t at position p is denoted by $t|_p$ and the result of replacing $t|_p$ with u at position p in t is denoted by $t[u]_p$. This notation is also used to indicate that u is a subterm of t. $\mathcal{V}(t)$ denotes the set of variables occurring in t. A term is *linear* if variables in $\mathcal{V}(t)$ occur at most once in t. A term is *ground* if $\mathcal{V}(t) = \emptyset$. Substitutions are written as in $\{X_1 \mapsto t_1, \ldots, X_n \mapsto t_n\}$ where t_i is assumed to be different from the variable X_i. We use Greek letters for substitutions and postfix notation for their application.

Definition 1 (Rewrite step). *Given a signature $\mathcal{F}$, a* term rewriting system *on $\mathcal{F}$ is a set of* rewrite rules $R = \{l_i \to r_i\}_{i \in I}$, *where* $l_i, r_i \in T(\mathcal{F}, \mathcal{X})$, $l_i \notin \mathcal{X}$, *and* $\mathcal{V}(r_i) \subseteq \mathcal{V}(l_i)$. *A term t* rewrites *to a term u at position p with the rule $l \to r$ and the substitution σ, written* $t \to_p^{l \to r} u$, *or simply* $t \to_R u$, *if* $t|_p = l\sigma$ *and* $u = t[r\sigma]_p$. *Such a term t is called* reducible. *Irreducible terms are said to be in* normal form.

We denote by $\to_R^+$ (resp. $\to_R^*$) the transitive (resp. transitive and reflexive) closure of the rewrite relation $\to_R$. The subindex R will be omitted when it is clear from the context.

Example 1. Consider a signature for lists of natural numbers, with function symbols z (with arity 0) and s (with arity 1) to build numbers; nil (with arity 0) to denote an empty list, cons (with arity 2) and append (with arity 2) to construct and concatenate lists, respectively, $\in$ (with arity 2) to test the membership of a natural number in a list. The list containing the numbers 0 and 1 is written then as cons(z, cons(s(z), nil)), or simply [z, s(z)] for short. We can specify list concatenation with the following rewrite rules: append(nil, X) $\to$ X and append(cons(Y, X), Z) $\to$ cons(Y, append(X, Z)). Then we have a reduction sequence: append(cons(z, nil), cons(s(z), nil)) $\to^*$ cons(z, cons(s(z), nil)).

Boolean operators, such as disjunction, conjunction, and a conditional, can be specified using a signature that includes the constants true and false. For example, conjunction is defined by the rules and(true, X) $\to$ X, and and(false, X) $\to$ false. The notation t_1 and $\ldots$ and t_n is syntactic sugar for and($\ldots$ and(and(t_1, t_2), t_3) $\ldots$), and *if b then s else t* is syntactic sugar for the term if-then-else(b, s, t), with the rewrite rules: if-then-else(true, X, Y) $\to$ X and if-then-else(false, X, Y) $\to$ Y.

For example, we can define the membership operator "$\in$" as follows:

$\in (X, \mathsf{nil}) \to \mathsf{false}$, $\in (X, \mathsf{cons}(H, L)) \to$ *if* $X = H$ *then* true *else* $\in (X, L)$, where we assume "$=$" is a syntactic equality test defined by standard rewrite rules. We will often write $\in$ as an infix operator.

Among the most important properties in term rewriting we have confluence and termination. A term rewriting system R is *confluent* if for all terms t, u, v: $t \to^* u$ and $t \to^* v$ implies $u \to^* s$ and $v \to^* s$, for some s; it is *terminating* if all reduction sequences are finite. If all left-hand sides of rules in R are linear and rules are non-overlapping (i.e., there are no superpositions of left-hand sides) then R is orthogonal. Orthogonality is a sufficient condition for confluence [16].

For the approach to distributed access control that we propose later, we use distributed term rewriting systems (DTRSs); DTRSs are term rewriting systems where rules are partitioned into modules, each associated with an identifier, and function symbols are annotated with such identifiers. We assume that each module has a unique identifier that is associated to the source of the definition of a function f (this can be a person, a site, ...). We say that a rule $f(t_1, \ldots, t_n) \to r$ defines f. There may be several rules defining f: for example, we may write f_ν to indicate that the definition of the function symbol f is stored in the site ν, where ν is a site identifier, or write f_u to indicate that the definition of the function symbol f is provided, or trusted, by the user u. If a symbol is used in a rule without a site annotation, we assume the function is defined locally. For more details on Distributed Term Rewriting Systems, we refer the reader to [9].

Fundamental Concepts of the Category-Based Access Control Meta-Model. We briefly describe below the key concepts underlying the category-based meta-model of access control. We refer the reader to [4] for a detailed description.

Informally, a category is any of several distinct classes or groups to which entities may be assigned. Entities are denoted uniquely by constants in a many sorted domain of discourse, including:

- A countable set $\mathcal{C}$ of categories, denoted $c_0, c_1, \ldots$
- A countable set $\mathcal{P}$ of principals, denoted $p_0, p_1, \ldots$
- A countable set $\mathcal{A}$ of named *actions*, denoted $a_0, a_1, \ldots$
- A countable set $\mathcal{R}$ of *resource identifiers*, denoted $r_0, r_1, \ldots$
- A finite set $\mathcal{A}uth$ of possible *answers* to access requests.

Additionally, the following sets are used in specific models:

- A countable set $\mathcal{S}$ of *situational identifiers*.
- A countable set $\mathcal{E}$ of *event identifiers*, denoted $e_0, e_1, \ldots$

We assume that principals that request access to resources are pre-authenticated. Situational identifiers are used to denote contextual or environmental information e.g., locations, times, system states, etc. The precise set $\mathcal{S}$ of situational identifiers that is admitted is application specific. Event identifiers uniquely denote happenings at a point in time. We adopt a one-dimensional, linear, discrete view of time, with a beginning and no end point. An important element in access control models is the request-response component. In the meta-model, the answer to a request may be one of a series of constants. For instance, the set $Auth$ might include {grant, deny, grant-if-obligation-is-satisfied, undefined}.

In addition to the different types of entities that we admit, we consider relationships between entities. The following relations are of primary importance for the specification of access control policies:

- *Principal-category assignment:* $\mathcal{PCA} \subseteq \mathcal{P} \times \mathcal{C}$, such that $(p, c) \in \mathcal{PCA}$ iff a principal $p \in \mathcal{P}$ is assigned to the category $c \in \mathcal{C}$.
- *Permissions:* $\mathcal{ARCA} \subseteq \mathcal{A} \times \mathcal{R} \times \mathcal{C}$, such that $(a, r, c) \in \mathcal{ARCA}$ iff the action $a \in \mathcal{A}$ on resource $r \in \mathcal{R}$ can be performed by principals assigned to the category $c \in \mathcal{C}$.
- *Authorisations:* $\mathcal{PAR} \subseteq \mathcal{P} \times \mathcal{A} \times \mathcal{R}$, such that $(p, a, r) \in \mathcal{PAR}$ iff a principal $p \in \mathcal{P}$ can perform the action $a \in \mathcal{A}$ on the resource $r \in \mathcal{R}$.

Thus, $\mathcal{PAR}$ defines the set of authorisations that hold according to an access control policy that specifies $\mathcal{PCA}$ and $\mathcal{ARCA}$.

Definition 2 (Axioms). *The relation $\mathcal{PAR}$ satisfies the following core axiom:*

$$(a1) \quad \forall p \in \mathcal{P}, \ \forall a \in \mathcal{A}, \ \forall r \in \mathcal{R}, \ \forall c \in \mathcal{C},$$
$$(p, c) \in \mathcal{PCA} \ \wedge \ (a, r, c) \in \mathcal{ARCA} \Rightarrow (p, a, r) \in \mathcal{PAR}$$

If an inclusion relationship between categories, written $c \subseteq c'$, is admitted then a more general version of the axiom for $\mathcal{PAR}$ can be defined thus:

$$(b1) \ \forall p \in \mathcal{P}, \ \forall a \in \mathcal{A}, \ \forall r \in \mathcal{R}, \ \forall c \in \mathcal{C},$$
$$(p, c) \in \mathcal{PCA} \wedge (\exists c' \in \mathcal{C}, c \subseteq c' \ \wedge (a, r, c') \in \mathcal{ARCA}) \Rightarrow (p, a, r) \in \mathcal{PAR}$$

In terms of principals, the semantics of $c \subseteq c'$ can simply be set inclusion (the set of principals assigned to $c \in \mathcal{C}$ is a subset of the set of principals assigned to $c' \in \mathcal{C}$), or an application specific relation may be used.

3   An Operational Specification of the Meta-model

The category-based meta-model of access control, henceforth denoted by $\mathcal{M}$, is based on a core axiom for $\mathcal{PAR}$ (see Def. 2). Operationally, this axiom can be realised through a set of function definitions, as we describe next.

3.1   Rewrite-Based Specification

Recall that authorisations, defined by the relation $\mathcal{PAR}$, are derived from $\mathcal{PCA}$ and $\mathcal{ARCA}$ (cf. Def. 2). The information contained in the relations $\mathcal{PCA}$ and $\mathcal{ARCA}$ will be modelled by the functions pca and arca, respectively. The function pca returns the list of categories assigned to a principal and arca returns a list of permissions assigned to a category. For instance, the following rewrite rule specifications can be used to define pca and arca: pca(p) $\rightarrow$ [c], arca(c) $\rightarrow$ $[(a_1, r_1), \ldots, (a_n, r_n)]$.

Definition 3. *The rewrite-based specification of the axiom* (a1) *in Def. 2 is given by the rewrite rule:*

(a2) $\mathsf{par}(P, A, R) \to if\ (A, R) \in \mathsf{arca}^*(\mathsf{pca}(P))\ then$ grant $else$ deny

where the function $\in$ *is a membership operator on lists (see Section 2),* grant *and* deny *are answers,[2] and* arca* *is a generalised version of the previously mentioned function* arca *to take into account lists of categories instead of a single category:*

$\mathsf{arca}^*(\mathsf{nil}) \to \mathsf{nil}$ $\mathsf{arca}^*(\mathsf{cons}(C, L)) \to \mathsf{append}(\mathsf{arca}(C), \mathsf{arca}^*(L))$

If we consider the more general axiom (b1) *in Def. 2, which involves an inclusion relationship between categories, the function* par *may be more generally defined as:*

(b2) $\mathsf{par}(P, A, R) \to if\ (A, R) \in \mathsf{arca}^*(\mathsf{contain}(\mathsf{pca}(P)))\ then$ grant $else$ deny

As the function name suggests, contain *computes the set of categories that contain any of the categories given in the list* $\mathsf{pca}(P)$. *For example, for a category* c, *this can be achieved by using a rewrite rule* $\mathsf{contain}([\mathsf{c}]) \to [\mathsf{c}, \mathsf{c}_1, \ldots, \mathsf{c}_n]$.

For optimisation purposes, one can compose the standard list concatenation operator append with a function removing the duplicate elements in the list.

An access request by a principal p to perform the action a on the resource r can then be evaluated simply by rewriting the term $\mathsf{par}(p, a, r)$ to normal form.

Proposition 1. *The rewrite based definition of* par *given in Def. 3 is a correct realisation of the axioms in Definition 2:* $\mathsf{par}(p, a, r) \to^*$ grant *if and only if* $(p, a, r) \in \mathcal{PAR}$.

Proof. Using the rules given in Definition 3, the normal form of $\mathsf{arca}^*(\mathsf{pca}(p))$ is a list containing the elements in $\mathsf{arca}(c)$ for each c in $\mathsf{pca}(p)$. Thus, if (a, r) is in $\mathsf{arca}(c)$ for some c in $\mathsf{pca}(p)$ then $\mathsf{par}(p, a, r) \to^*$ grant. By assumption, $(a, r) \in \mathsf{arca}(c)$ iff $(a, r, c) \in \mathcal{ARCA}$ and $c \in \mathsf{pca}(p)$ iff $(p, c) \in \mathcal{PCA}$. Therefore $\mathsf{par}(p, a, r) \to^*$ grant if $(a, r, c) \in \mathcal{ARCA}$ and $(p, c) \in \mathcal{PCA}$, as expected.

A range of access control concepts can be understood in category-based terms. For example, notions like provisional authorisations can be represented in category-based terms by defining a specific category c of principals that assume an obligation, together with the corresponding arca definition for c. By choosing different notions of category and corresponding definitions of pca, contain and arca, the core elements of $\mathcal{M}$ can be specialised in multiple ways to define different instances of the meta-model. The rewrite rules may contain algebraic and arithmetic operators as well as functions that are application-specific. Moreover,

---

[2] Answers different from grant or deny can also be accommodated in the definition (for instance, an undefined answer could be returned when the pair (A, R) is not in the list of permissions for the category of P).

site annotations (as defined in Section 2) can be used for functions defined at a remotely accessible source.

We demonstrate the expressiveness of our proposal through an example that shows how complex access control requirements can be simply represented in terms of $\mathcal{M}$.

Example 2. Consider the following access control requirements:

> Employees in a company are classified as Managers (Manager), Senior Managers (Senior_mng) or Senior Executives (Senior_exec). Any principal that is a Senior Executive (i.e., a member of the category Senior_exec) is permitted to read the personal information, including salary, stored in an employee's file, provided the employee works in a profitable branch and is categorised as a Manager. To be categorised as a Senior Executive, a principal must be categorised as being a senior manager (Senior_mng) (according to the source of this information, v_1) and must be a member of the executive board of the company. All managers' names are recorded locally, and the list of profitable branches is kept up to date at site v_2.

Thus, to grant the permission to read a principal's file, considered as a resource in this example, it is necessary to deal with various forms of categorisation including categorisation of an institution (a branch office) having a particular status (of being profitable).

This policy can be represented by using the following specialised rules of $\mathcal{M}$ together with the rules that define pca remotely, at v_1.

$$\begin{aligned} \mathsf{pca}(P) \rightarrow\ &if\ \mathsf{Senior\_mng} \in \mathsf{pca}_{v1}(P) \\ &then\ (if\ P \in \mathsf{Exec\_board}\ then\ [\mathsf{Senior\_exec}] \\ &\qquad else\ [\mathsf{Senior\_mng}]) \\ &else\ [\mathsf{Manager}] \end{aligned}$$

The specification of the arca function is given below. The auxiliary function profbranch, defined at site v_2, returns the list of branches that are profitable. The local function manager returns the name of the manager of a branch B given as a parameter, and managers(LB) does the same for a list of branches. We omit the definition of the function zip-read, which, given a list $L = [l_1, \ldots, l_n]$, returns a list of pairs of the form $[(\mathsf{read}, l_1), \ldots, (\mathsf{read}, l_n)]$.

$$\begin{aligned} \mathsf{arca}(\mathsf{Senior\_exec}) &\rightarrow \mathsf{zip\text{-}read}(\mathsf{managers}(\mathsf{profbranch}_{v2})) \\ \mathsf{managers}(\mathsf{nil}) &\rightarrow \mathsf{nil} \\ \mathsf{managers}(\mathsf{cons}(B, LB)) &\rightarrow \mathsf{cons}(\mathsf{manager}(B), \mathsf{managers}(LB)) \\ \mathsf{Exec\_board} &\rightarrow [(p_1, \ldots, p_n)] \\ \mathsf{profbranch}_{v2} &\rightarrow [(b_1, \ldots, b_m)] \end{aligned}$$

For more expressive power, categories can be parameterised. For instance, the category *client* can be parameterised in such a way that *client*(b) indicates a client of the branch b. More generally, categories can be represented by terms instead of simply constants, and variables may be used in parameterised expressions.

3.2 Policy Analysis: Proving Properties of Policies

Specifying access control policies via term rewriting systems, which have a formal semantics, has the advantage that this representation admits the possibility of proving properties of policies, and this is essential for policy acceptability [21]. Rewriting properties like confluence (which implies the unicity of normal forms) and termination (which implies the existence of normal forms for all terms) may be used to demonstrate satisfaction of essential properties of policies, such as consistency.

More specifically, we are interested in the following properties of access control policies:

Totality: Each access request from a valid principal p to perform a valid action a on a resource r receives an answer (e.g., grant, deny, undeterminate).

Consistency: For any $p \in \mathcal{P}$, $a \in \mathcal{A}$, $r \in \mathcal{R}$, at most one result is possible for an authorisation request $\mathsf{par}(p, a, r)$.

Soundness and Completeness: For any $p \in \mathcal{P}$, $a \in \mathcal{A}$, $r \in \mathcal{R}$, an access request by p to perform the action a on r is granted if and only if p belongs to a category that has the permission (a, r).

Totality and consistency can be proved, for policies defined as term rewriting systems, by checking that the rewrite relation is confluent and terminating. Termination ensures that all access requests produce a result (e.g. a result that is not grant or deny is interpreted as undeterminate) and confluence ensures that this result is unique. The soundness and completeness of a policy can be checked by analysing the normal forms of access requests.

Confluence and termination of rewriting are undecidable properties in general, but there are several results available that provide sufficient conditions for these properties to hold, we give examples below.

Termination. An access control model defined as an instance of our meta-model provides a notion of category and the corresponding definitions of the relations between categories, between principal and categories and between categories and permissions. An access control policy can then be defined in terms of the chosen access control model. The full definition of the policy can be seen as a *hierarchical* rewrite system, where the basis includes the set of constants identifying the main entities in the model (e.g., principals, categories, etc.) as well as the set of auxiliary data structures (such as booleans, lists) and functions on these data structures. The next level in the hierarchy contains the parameter functions of the model, namely pca, arca, arca*, contain. Finally the last level of the hierarchy consists of the definition of the function par.

Several sufficient conditions for termination of rewrite systems defined as a hierarchical union of rules are available. For instance, a hierarchical term rewriting system is terminating if the basis of the hierarchy is terminating and non-duplicating (i.e., rules do not duplicate variables in the right-hand side) and in the next levels of the hierarchy the recursive functions are defined by rules that satisfy a general scheme of recursion, where recursive calls on the right-hand sides of rules are made on subterms of the left-hand side [12]. Thus, for the examples in this paper, provided the auxiliary functions in the basis of the hierarchy are defined by non-duplicating and terminating rewrite rules, the system is terminating (notice that the function par is not recursive). We discuss an example in more detail below.

Confluence. Confluence can be proved in various ways. Orthogonal systems are confluent, as shown by Klop [16]. A less restrictive condition, for systems that terminate, is the absence of critical pairs (the latter, combined with termination implies confluence by Newman's lemma [20]).

Application. To illustrate the methodology, we can prove for instance the consistency and totality of the policy specified in Example 2.

Property 1. The policy in Example 2 is consistent and total.

Proof.

- Consistency: It is not difficult to see that the rules used in Example 2 form an orthogonal (hence confluent) system, therefore the policy is consistent.
- Totality: As mentioned above, we can see the policy as a hierarchical system. The basis contains the rules defining the functions Exec_board, profbranch, manager, managers, zip-read, together with rules for booleans and lists (as defined in Example 1), which are non-duplicating and terminating. The next levels in the hierarchy define functions that satisfy the general scheme of recursion. Therefore, the whole policy seen as a hierarchical union of rewrite systems is terminating [12]; hence the policy is total.

Concerning the soundness and completeness of the policy described in Example 2, we can prove that if p is a Senior Executive and m is the manager m of a profitable branch b, then an access request from p to read m's file will be granted. For this, we analyse the normal forms of access requests. More precisely: if p is a Senior Executive and m is the manager of a profitable branch, as recorded in v_2, then the normal form of $\mathsf{par}(p, \mathsf{read}, m)$ is grant, since $\mathsf{par}(p, \mathsf{read}, m) \to^* \mathsf{grant}$.

4   Expressive Power: Access Control Models in $\mathcal{M}$

In this section we show how a range of existing access control models, as well as novel models of access control, can be represented as specialised instances of our meta-model. We specify traditional (static) access control models, such as RBAC, DAC and MAC (including the well-known Bell-LaPadula model), as well as dynamic models, such as DEBAC. We also discuss briefly the representation of trust-based access control models.

4.1   Hierarchical Role Based Access Control

Standard Role-Based Access Control (RBAC) models [13,2] assume a single (limited) form of category: the role. In ANSI Hierarchical RBAC, role hierarchies are the only form of category-category relationships that are admitted: a role hierarchy is a partial ordering of roles by seniority, where a senior role inherits the privileges of its subordinate roles [2]. In an RBAC policy, each principal is assigned to one or several roles, and permissions are associated to roles. Briefly, RBAC models can be characterised as follows:

A request $access(p, a, r)$ from a principal p to perform the action a on the resource r results in grant *or* deny *depending on whether the permission (a, r) is associated to p's role (or its subordinate roles).*

In terms of our access control primitives, hierarchical RBAC can be expressed using Def. 3 (b2) where categories are limited to roles. Thus, $\mathsf{pca}(p) \rightarrow [r_1, \ldots, r_n]$, where $r_1, \ldots, r_n$ are the roles of p. We use an additional function dc in the definition of $\mathsf{contain}$ where $\mathsf{dc}(r_i) = [r_1, \ldots, r_j]$ means that $r_1, \ldots r_j \in \mathcal{C}$ are direct subordinate roles of $r_i \in \mathcal{C}$ (hence r_i is directly senior to $r_1 \ldots r_j$). In order to compute the hierarchy of roles, we can use the following rewrite rules:

$$\mathsf{contain(nil)} \rightarrow \mathsf{nil}$$
$$\mathsf{contain(cons}(C, L)) \rightarrow \mathsf{append(cons}(C, \mathsf{contain(dc}(C))), \mathsf{contain}(L))$$

The permissions assigned to a role, which includes its permissions and the permissions of its direct subordinate roles, are then computed by the function arca^*, as already defined in Section 4.1. We can then evaluate access requests by rewriting: the answer for a request by p to perform the action a in r can be computed by rewriting the term $\mathsf{par}(p, a, r)$ to normal form, using rule (b2).

Property 2 (Expressing RBAC in $\mathcal{M}$). Given an RBAC policy, an access request by a principal p to perform the action a in the resource r is granted (respectively denied) if $\mathsf{par}(p, a, r) \rightarrow^*$ grant (respectively $\mathsf{par}(p, a, r) \rightarrow^*$ deny).

Proof. In accordance with the characterisation of RBAC given above, to compute an answer to the access request we need to check whether (a, r) is in the list of permissions returned by the function arca^*, whose argument is the list of roles of p as computed by $\mathsf{pca}(p)$ (and the function $\mathsf{contain}$ in case of a hierarchy of roles).

Thus, RBAC and Hierarchical RBAC can be obtained as instances of the meta-model $\mathcal{M}$. We discuss below an extension of RBAC that takes into account time and location constraints. More general models, such as *Event-based Access Control (DEBAC)* models [9], will be discussed in Section 4.2.

Temporal and Location-Based Models. To satisfy a wider range of application-specific requirements, constants can be introduced to express intervals of time (i.e., $Int(T_{start}, T_{stop})$) during which principal-category assignments and permission-category assignments hold.

To accommodate such generalisations, we consider Def. 3 (b2) and we specialise the rewrite rules for arca and pca. For example, to constrain assignment of a principal p to a category by time (i.e., to help satisfy the Principle of Least Privilege) we may use a rule of the following form:

$$\mathsf{pca}(p) \rightarrow if \text{ current\_time} \in \mathsf{Int}(8am, 1pm) \text{ } then \text{ } [\mathsf{clerk}, \mathsf{record\_keeper}]$$
$$else \text{ } if \text{ current\_time} \in \mathsf{Int}(1pm, 6pm) \text{ } and \text{ delegated} \in \mathsf{pca}_b(p)$$
$$then \text{ } [\mathsf{boss\_assistant}] \text{ } else \text{ } [\mathsf{out\_of\_office}]$$

Supposing only the *boss assistant* has the permission of reading the emails of his boss, i.e. $\mathsf{arca}(\mathsf{boss\_assistant}) \rightarrow [(\mathsf{read}, \mathsf{mailBoss})]$, a request of the form $\mathsf{par}(p, \mathsf{read}, \mathsf{mailBoss})$ made in the morning will lead to a denial of access:

$$\mathsf{par}(p, \mathsf{read}, \mathsf{mailBoss}) \rightarrow^*$$
$$if \text{ } (\mathsf{read}, \mathsf{mailBoss}) \in \mathsf{arca}^*([\mathsf{clerk}, \mathsf{record\_keeper}]) \text{ } then \text{ grant } else \text{ deny} \rightarrow^* \text{deny}$$

but the same request made at $2pm$ will be successful, if p has been delegated by the boss, as recorded in $\mathsf{pca}_b(p)$.

Concerning location-based models, suppose that a categorisation of principals were required that took into account their current location to determine the principal's set of authorisations. A contain relation may then be defined in terms of the function dc where dc is used to define geographical regions that are ordered by direct containment, e.g., $\mathsf{dc}(europe) = [Uk, Fr, It, \ldots]$). In this case, a category c associated to a principal p represents a categorisation of p by location.

Mandatory Access Control and Discretionary Access Control. MAC and DAC can be viewed as special cases of RBAC (as has already been noted in the access control literature e.g., [22]). In terms of our proposed framework, a version of the Bell-LaPadula model may be viewed as a restricted form of the Hierarchical RBAC model, with the second parameter of par limited to the read or write action. In this case, the containment relationship is an ordering of categories that are restricted to being defined on a common set of security classifications for resources and security clearances for principals.

The function contain is defined as in the RBAC instance and returns a list with the principal's category as head of the list and all of its subordinate categories as tail of the list: $\mathsf{contain}(c) \rightarrow [c, c_1, \ldots, c_n]$. The function par is specified as in Def. 3 (b2), using $\mathsf{arca}^*(\mathsf{contain}(\mathsf{pca}(p)))$. We specify two kinds of permissions: we define a function $\mathsf{arcaw}(c)$ returning the write privileges (write, r_i) associated to a category c, and a function $\mathsf{arcar}(c)$ returning the read privileges (read, r_i) associated to a category c. Then we can express the requirements "write only at the subject's classification level" and "no read up" (which are core axioms of strict MAC, as the latter term is interpreted in Bell-LaPadula terms) by using the following definition of function arca^*:

$$\mathsf{arca}^*(\mathsf{nil}) \rightarrow \mathsf{nil} \quad \mathsf{arca}^*(\mathsf{cons}(C, L)) \rightarrow \mathsf{cons}(\mathsf{arcaw}(C), \mathsf{arcar}^*(\mathsf{cons}(C, L)))$$
$$\mathsf{arcar}^*(\mathsf{cons}(C, L)) \rightarrow \mathsf{cons}(\mathsf{arcar}(C), \mathsf{arcar}^*(L))$$

Thus, when evaluating $\mathsf{arca}^*(\mathsf{contain}(\mathsf{pca}(p)))$, the arcaw function is applied at the category level, while arcar is recursively called on all the subordinates of the initial category.

Property 3 (Expressing Bell-LaPadula in $\mathcal{M}$). An access request $access(p, read, r)$ or $access(p, write, r)$ in a Bell-LaPadula policy is granted (respectively denied) iff $par(p, read, r) \rightarrow^*$ grant (respectively deny) or $par(p, write, r) \rightarrow^*$ grant (respectively deny).

Proof. This can be shown using the fact that the policy can be expressed as a special case of a RBAC policy (as detailed e.g. in [22]).

We also note that the discretionary access control model of Unix can be naturally represented, with Unix users being viewed as particular types of principals and with *group* and *other* being a restricted set of categories of principals.

4.2   Generalising RBAC: Event-Based Access Control (DEBAC)

In the DEBAC model [9], users may request to perform actions on resources that are potentially accessible from any site in a distributed system. A user is assigned to a particular category on the basis of the history of events that relate to the user. DEBAC generalises RBAC by taking into account the actions that involve a user in order to compute its category. Briefly, DEBAC can be described as follows:

A user is permitted to perform an action a on a resource r if and only if the user is assigned to a category c to which a access on r has been assigned, according to the history of events.

To model the history of an agent's actions, the notion of an event is extensively used in DEBAC. We view events as structured and described via a sequence h of ground terms of the form event(e_i, p, a, t) where event is a data constructor of arity four, e_i $(i \in \mathbb{N})$ denotes unique event identifiers, p identifies a principal, a is an action associated to the event, and t is the time when the event happened.

Example 3. Consider a university where students may acquire permissions (by changing category) as they pass their exams. The function pca is defined as follows:

$$\text{pca}(P) \rightarrow \text{head}(\text{action}(P, \text{h})))$$

where action is a function returning the categories associated to a principal, according to the events in which the principal is involved (stored in the list h). These functions could be defined as follows:

$$\text{action}(P, \text{nil}) \quad\quad \rightarrow [\text{Student}]$$
$$\text{action}(P, \text{cons}(E, h)) \rightarrow if\ \text{User}(E) = P\ then\ \text{cons}(\text{categ}(E), \text{action}(P, h))$$
$$\quad\quad\quad else\ \text{action}(P, h)$$

The function action looks recursively in the list of events, and uses an application-specific function categ that associates a category to a principal according to the particular event in which he/she was involved. For instance, in the university context we may have

$$
\begin{array}{ll}
\mathsf{categ}(\mathsf{event}(E, P, enroll, T)) & \rightarrow [\mathsf{Registered}] \\
\mathsf{categ}(\mathsf{event}((E, P, pay, T)) & \rightarrow [\mathsf{Regular}] \\
\mathsf{categ}(\mathsf{event}(E, P, exampassed, T)) & \rightarrow if\ \mathsf{pass}(P, exams1styear) \\
& \quad then\ [\mathsf{2nd\_year\ Student}]\ else\ [\mathsf{Irregular}]
\end{array}
$$

In this case a student, according to the actions he/she has performed may be assigned for instance to a *2nd_year Student* category.

Property 4 (Expressing DEBAC in $\mathcal{M}$). An access request $access(p, a, r)$ in a DEBAC policy is granted (respectively denied) iff $\mathsf{par}(P, A, R) \rightarrow^*$ grant (respectively $\mathsf{par}(P, A, R) \rightarrow^*$ deny).

Proof. To compute an answer to the access request in accordance with the given DEBAC's definition, in our specification the main function par checks whether (a, r) is in the list of permissions returned by the function arca*, whose argument is the list of categories of p as computed by $\mathsf{pca}(p)$. Thus, provided the function categ is specified for any event of the list, p's category is computed by $\mathsf{pca}(p)$ which makes use of the function action to go through the history of events.

Generalising DEBAC in $\mathcal{M}$. We can define a generalisation of the DEBAC model in $\mathcal{M}$ by considering categorisation depending not only on the history of an agent's actions, but also on events related to other agents in the system. Moreover we can introduce an "ascribed" status associated to the user (of which a role assignment is a particular type) and combine it with a status resulting from his/her actions to compute the overall status of a user (see Example 4). This overall categorisation is used as the basis for determining authorised forms of actions. Ascription and action are simply particular types of category; as such, two categories can be combined for access control decisions to be made.

Example 4. Consider a bank scenario where two clients share a joint account. In this case, actions of one client on the joint account, such as withdrawals or deposits, may affect the category (hence the access permissions) of the second client. We use the following *pca* specification to compute the combination of the ascribed status associated to a principal and its action status.

$$
\begin{array}{ll}
\mathsf{pca}(P) & \rightarrow \mathsf{cons}(\mathsf{ascribed}(P), \mathsf{action}(P, \mathsf{h})) \\
\mathsf{ascribed}(P) & \rightarrow if\ \mathsf{age}(P) \geq 60\ then\ [\mathsf{Senior}]\ else\ [\mathsf{Junior}] \\
\mathsf{action}(P, \mathsf{nil}) & \rightarrow [\mathsf{Client}] \\
\mathsf{action}(P, \mathsf{cons}(E, h)) & \rightarrow \mathsf{cons}(\mathsf{categ}(P, E), \mathsf{action}(P, h))
\end{array}
$$

with the following rule associated to the event of a withdrawal:

$$
\begin{array}{l}
\mathsf{categ}(P, \mathsf{event}(E, P', withdrw500, T)) \rightarrow if\ \mathsf{jointaccount}(P, P')\ then \\
\quad if\ \mathsf{balance}(\mathsf{account}(P')) \leq 0\ then\ [\mathsf{BlackList}]\ else\ [\mathsf{Normal}]
\end{array}
$$

Thus, according to his/her age and the actions he/she has performed, as well as the actions of a partner on the joint account, the client may be assigned for instance to a *Senior BlackList* category.

It is also possible to introduce in the DEBAC model a function for verifying the validity of a status assignment over time, and use it for instance to specify the Chinese Wall policy, as shown in Example 5.

Example 5 (Chinese Wall policy). In the Chinese Wall model resources are grouped into "conflict of interest classes" and by mandatory ruling all principals are allowed access to at most one resource belonging to each such conflict of interest class. This model can be easily specified in terms of categories in the meta-model $\mathcal{M}$. It is sufficient to adapt the DEBAC specification shown previously. We assume that each category $c_0, \ldots c_n$ corresponds to an interest class and that by default all principals are assigned to all categories, i.e. $\mathsf{action}(P, \mathsf{nil}) \to [c_0, \ldots c_n]$. In order to verify the validity of a status assignment over time, we introduce the function filter. This function checks that the occurrence of a particular event in the history does not create conflicts with some previous category assignments (if it does, those categories are not included in the list computed by the function action).

$$
\begin{aligned}
\mathsf{action}(P, \mathsf{cons}(E, h)) \to \; &if \; \mathsf{User}(E) = P \\
&then \; \mathsf{cons}(\mathsf{categ}(E), \mathsf{filter}(E, \mathsf{action}(P, h))) \\
&else \; \mathsf{action}(P, h) \\
\mathsf{filter}(E, \mathsf{nil}) \qquad \to \; &\mathsf{nil} \\
\mathsf{filter}(E, \mathsf{cons}(C, h)) \to \; &if \; \mathsf{conflict}(E, C) \; then \; \mathsf{filter}(E, h) \\
&else \; \mathsf{cons}(C, \mathsf{filter}(E, h))
\end{aligned}
$$

In other words, when a principal obtains access to a resource belonging to a particular class, conflicts between this event and the list of categories are checked and the access rights of the principal are updated.

4.3   Trust-Based Models

As far as access control models based on trust are concerned, we regard the association of a trust measure with a principal as the assignment of the principal to a category of users that have the same degree of trust according to some authority. In particular, if a policy author α asserts that a principal p is assigned to a category c, then any policy author that sufficiently trusts α's categorisation of p can refer to that category in its specifications of access control requirements. The next example that we give illustrates the representation of trusted third-party assertions in our approach.

Example 6 (Third-party assertions). Consider a local policy specification, with the following access control requirements.

> Any principal P is assigned to the category approved_uni if a known principal u is assigned to the category of trusted_on_uni in the database d of a trusted source, and u asserts that P is assigned to the category good_university.

To represent these access control requirements, the following definition of pca is sufficient:

$$\mathsf{pca}(P) \rightarrow if \ \mathsf{trusted\_on\_uni} \in \mathsf{pca_d}(u) \ and \ \mathsf{good\_university} \in \mathsf{pca_u}(P)$$
$$then \ [\mathsf{approved\_uni}]$$

To appreciate how $\mathcal{M}$ can be specialised in the context of the "policy aware web" [24], consider for example a *mutual access partnership (MAP)*. In such a policy specification, two principals that trust each other allow the partner the rights for accessing their resources. In other words, if $p' \in \mathcal{P}$ and $p'' \in \mathcal{P}$ are in a MAP then p' will allow p'' to access its resources and conversely. To represent the MAP model in $\mathcal{M}$, it is sufficient for a policy author to declare a category for each pair of principals in a MAP.

Note that the contain relationship between categories can be arbitrarily defined; we could also specify an equivalence relation between categories (such relationships are often useful in trust-based models [19]). For instance, consider an access control model where if members of category c_2 trust the assertions of an immediately "superior" authority category c_1 and members of category c_3 similarly trust c_1, then c_2 and c_3 trust each others' assertions. The (Euclidean) relationship that is required in this case can simply be captured by defining an equivalence between c_2 and c_3.

$$\mathsf{contain}(c_1) \rightarrow [c_1, c_2, c_3] \quad \mathsf{contain}(c_2) \rightarrow [c_2, c_3] \quad \mathsf{contain}(c_3) \rightarrow [c_2, c_3]$$

With the specification given above, access requests from members of c_2 or c_3 will be treated in the same way. Principals in category c_1 have all the permissions of c_2 and c_3, reflecting the fact that c_2 and c_3 trust c_1.

5   Related Work

Term rewriting has been used to model a variety of problems in security, from the analysis of security protocols [6] to the definition of policies for controlling information leakage [11]. On access control specifically, Koch et al. [17] use graph transformation rules to formalise RBAC, and more recently, [5,23,9] use term rewrite rules to model particular access control models and to express access control policies. The approach used in [5,23,9] is operational: access control policies are specified as sets of rewrite rules and access requests are specified as terms that are evaluated using the rewrite rules. Our work addresses similar issues to [5,17,23,9], but is based on a notion of a meta-model of access control from which various models can be derived, instead of formalising a specific kind of model such as RBAC or DEBAC.

Several proposals for general models and languages for access control have already been described in the literature, but neither of them provides the level of generality of the category-based approach. For example, the Generalised TRBAC model [15] and ASL [14] aim at providing a general framework for the definition of policies. In GTRBAC, the focus is on the notion of a role (interpreted as being synonymous with the notion of job function). In ASL, users, groups and roles are admitted in the language.

Abadi et al [1] ABLP logic also provides a formal framework for reasoning about a wide range of features of access control. The focus in ABLP logic is on language constructs for formulating access control policies and axioms and inference rules for defining a system for proof, e.g., for proving authorised forms of access. In contrast, the category-based approach is based on abstracting, from the generalities of access control models, the common aspects of access control from which core functions are identified. The rewrite-based specification of the meta-model is well-adapted to the evaluation of access requests, and emphasises the use of rewriting properties to derive properties of the policies.

Li et al.'s *RT* family of role-trust models [18] provides a general framework for defining access control policies, which can be specialised for defining specific policy requirements (in terms of credentials). The category-based meta-model, however, can be instantiated to include concepts like times, events, actions and histories that may be used to specify principal-category assignments, but which are not included as elements of *RT*.

6 Conclusions and Further Work

We have given a rewrite-based specification of a meta-model of access control that is based on common, core concepts of access control models. The term rewriting approach can be used to give a meaningful semantics to policies in the case of both centralised and distributed computer systems. Rewrite rules provide a declarative specification of access control policies which facilitates the task of proving properties of policies. Also, term rewriting rules provide an executable specification of the access control policy. In future work, we will investigate the design of languages for policy specification and the practical implementation of category-based policies; in particular, an access control policy may be transformed into a MAUDE [10] program by adding type declarations for the function symbols and variables used and by making minor syntactical changes (see [8]).

References

1. Abadi, M., Burrows, M., Lampson, B.W., Plotkin, G.D.: A calculus for access control in distributed systems. ACM Trans. Program. Lang. Syst. 15(4), 706–734 (1993)
2. ANSI. RBAC, INCITS 359-2004 (2004)
3. Baader, F., Nipkow, T.: Term rewriting and all that. Cambridge University Press, Great Britain (1998)
4. Barker, S.: The next 700 access control models or a unifying meta-model? In: Proceedings of the ACM Int. Conf. SACMAT 2009, pp. 187–196. ACM Press, New York (2009)
5. Barker, S., Fernández, M.: Term rewriting for access control. In: Damiani, E., Liu, P. (eds.) Data and Applications Security 2006. LNCS, vol. 4127, pp. 179–193. Springer, Heidelberg (2006)

6. Barthe, G., Dufay, G., Huisman, M., Melo de Sousa, S.: Jakarta: a toolset to reason about the JavaCard platform. In: Attali, S., Jensen, T. (eds.) E-smart 2001. LNCS, vol. 2140, p. 2. Springer, Heidelberg (2001)
7. Bell, D.E., LaPadula, L.J.: Secure computer system: Unified exposition and multics interpretation. MITRE-2997 (1976)
8. Bertolissi, C., Fernández, M.: Time and location based services with access control. In: Proceedings of the 2nd IFIP International Conference on New Technologies, Mobility and Security. IEEEXplore (2008)
9. Bertolissi, C., Fernández, M., Barker, S.: Dynamic event-based access control as term rewriting. In: Barker, S., Ahn, G.-J. (eds.) Data and Applications Security 2007. LNCS, vol. 4602, pp. 195–210. Springer, Heidelberg (2007)
10. Clavel, M., Durán, F., Eker, S., Lincoln, P., Martí-Oliet, N., Meseguer, J., Talcott, C.: The Maude 2.0 system. In: Nieuwenhuis, R. (ed.) RTA 2003. LNCS, vol. 2706, pp. 76–87. Springer, Heidelberg (2003)
11. Echahed, R., Prost, F.: Security policy in a declarative style. In: Proc. 7th ACM-SIGPLAN Symposium on Principles and Practice of Declarative Programming (PPDP 2005). ACM Press, New York (2005)
12. Fernández, M., Jouannaud, J.-P.: Modular termination of term rewriting systems revisited. In: Reggio, G., Astesiano, E., Tarlecki, A. (eds.) Abstract Data Types 1994 and COMPASS 1994. LNCS, vol. 906. Springer, Heidelberg (1995)
13. Ferraiolo, D.F., Sandhu, R.S., Gavrila, S.I., Richard Kuhn, D., Chandramouli, R.: Proposed NIST standard for role-based access control. ACM TISSEC 4(3), 224–274 (2001)
14. Jajodia, S., Samarati, P., Sapino, M., Subrahmaninan, V.S.: Flexible support for multiple access control policies. ACM TODS 26(2), 214–260 (2001)
15. Joshi, J., Bertino, E., Latif, U., Ghafoor, A.: A generalized temporal role-based access control model. IEEE Trans. Knowl. Data Eng. 17(1), 4–23 (2005)
16. Klop, J.-W., van Oostrom, V., van Raamsdonk, F.: Combinatory reduction systems, introduction and survey. Theoretical Computer Science 121, 279–308 (1993)
17. Koch, M., Mancini, L., Parisi-Presicce, F.: A graph based formalism for rbac. In: SACMAT, pp. 129–187 (2004)
18. Li, N., Mitchell, J.C., Winsborough, W.H.: Design of a role-based trust-management framework. In: IEEE Symposium on Security and Privacy, pp. 114–130 (2002)
19. Liau, C.-J.: Belief, information acquisition, and trust in multi-agent systems–a modal logic formulation. Artif. Intell. 149(1), 31–60 (2003)
20. Newman, M.H.A.: On theories with a combinatorial definition of equivalence. Annals of Mathematics 43(2), 223–243 (1942)
21. Department of Defense. Trusted computer system evaluation criteria (1983); DoD 5200.28-STD
22. Sandhu, R.S., Munawer, Q.: How to do discretionary access control using roles. In: ACM Workshop on Role-Based Access Control, pp. 47–54 (1998)
23. Santana de Oliveira, A.: Rewriting-based access control policies. In: Proceedings of SECRET 2006, Venice, Italy. Electronic Notes in Theoretical Computer Science. Elsevier, Amsterdam (2007) (to appear)
24. Weitzner, D.J., Hendler, J., Berners-Lee, T., Connolly, D.: Creating a policy-aware web: Discretionary, rule-based access for the world wide web. In: Web and Information Security (2006)

Idea: Efficient Evaluation of Access Control Constraints

Achim D. Brucker and Helmut Petritsch

SAP Research, Vincenz-Priessnitz-Str. 1, 76131 Karlsruhe, Germany
{achim.brucker,helmut.petritsch}@sap.com

Abstract. Business requirements for modern enterprise systems usually comprise a variety of dynamic constraints, i. e., constraints that require a complex set of context information only available at runtime. Thus, the efficient evaluation of dynamic constraints, e. g., expressing separation of duties requirements, becomes an important factor for the overall performance of the access control enforcement.

In distributed systems, e. g., based on the service-oriented architecture (SOA), the time for evaluating access control constraints depends significantly on the protocol between the central Policy Decision Point (PDP) and the distributed Policy Enforcement Points (PEPs).

In this paper, we present a policy-driven approach for generating customized protocol for the communication between the PDP and the PEPs. We provide a detailed comparison of several approaches for querying context information during the evaluation of access control constraints.

Keywords: distributed policy enforcement, XACML, access control.

1 Introduction

Business regulations, e. g., Basel II [3] or the Sarbanes-Oxley Act [16], often require the enforcement of dynamic or context-aware access control policies [9, 17], for example, (dynamic) separation-of-duties. As many widely used policy languages, e. g., [6, 15], are not supporting such context requirements as first class citizens, they are usually encoded as access control constraints [7]. For example, XACML [15], PERMIS [6], or SecureUML [4] are supporting access control constraints. By definition, these constraints depend on context information such as attributes of a resource (e. g., its owner, last modification date, shipping destination) and, as such, are not amenable for caching access control decisions [12]. Thus, an efficient attribute retrieval becomes even more important; astonishingly this is left unspecified in many policy frameworks, e. g., XACML [1] or EPAL [2].

In distributed systems following the service-oriented architecture (SOA), business services are provided by orchestrating a set of loosely coupled technical services. Still, authentication and authorization rely on centralized components: single sign on implementations authenticate users within highly distributed systems using a centralized authority. Moreover, while access control policies are enforced by distributed Policy Enforcement Points (PEPs), the central Policy

F. Massacci, D. Wallach, and N. Zannone (Eds.): ESSoS 2010, LNCS 5965, pp. 157–165, 2010.
© Springer-Verlag Berlin Heidelberg 2010

Decision Point (PDP) evaluates the underlying access control requests. Such a centrally managed and administered PDP stores access control policies for all secured services. For evaluation of dynamic constraints, attributes need to be resolved (i. e., accessible) at runtime within the PDP. Albeit, often the resolution of attributes can only be done within the service (i. e., its PEP) requesting the policy evaluation. Thus, resolving attributes often requires substantial network communication between the centralized PDP and the various distributed PEPs. All PDP implementations we are aware of are based on an iterative approach. In contrast, we propose to optimize the required communication overhead by using a static policy analysis that allows for sending all attributes required for a specific access control request at once.

Our contributions are three-fold: first, we present an approach for pre-computing the required attributes for evaluating access control constraints, second, we analyze and present the performance tests for several attribute resolution strategies for distributed enterprise systems, and, third, we give guidance on choosing an optimal resolution strategy based on multiple criteria.

2   Efficient Attribute Resolution

2.1   Context Attributes

Context attributes can be classified with respect to the runtime environment in which they can be resolved efficiently. For example, while information about the accessed object is usually only available in the context of the service issuing the access control request, information about a user (e. g., his roles) are often only available within the PDP. Thus, attributes can be categorized into:

PDP attributes are only available within the centralized security infrastructure, i. e., the PDP. The information about role hierarchy membership is, usually, an example for this kind of information.

Service attributes are only available within the client application or service, e. g., owner of a resource, specific attributes of a resource (e. g., balance of a bank account) or number of threads running on the application server.

Global attributes are equally resolvable from either the PDP or the services. For example, this is the case for attributes that need to be resolved by an additional service, e. g., the single sign-on or identity provider.

While in common implementations such a classification is left implicit, we propose to use such a classification explicitly.

2.2   Attribute Resolution Strategies

There exist several approaches for the resolution of context attributes during the evaluation of access control requests. Fig. 1 illustrates the approaches we will discuss in the following. These sequence diagrams only illustrate the resolution of service attributes, i. e., within the PEP. The resolution of global or PDP attributes using a Policy Information Point (PIP) or context provider is left out.

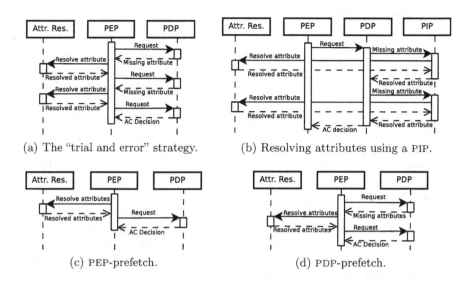

Fig. 1. Different strategies for resolving context attributes

Trial and error: The PDP requests the required attributes from the PEP using an iterative approach (see Fig. 1a). If the evaluation of an access control request requires a service attribute, the PDP returns a missing attribute message back to the client. As the PDP is stateless, the initial request has to be re-send by the PEP until all required attributes are supplied.

PIP: The PIP is responsible for resolving all types of attributes (see Fig. 1b). For example, the PIP queries the service or PEP over an additional Web service interface, which requires this interface be reachable from the PIP.

For increasing the performance of access control requiring the resolution of context attributes, we propose to pre-compute the set of required attributes using a static analysis of the policy together with one of the following two strategies:

PEP Prefetch: Using a pre-computed look-up-table mapping access control request to the (potentially over-approximated) set of required service attributes, the PEP can easily determine which attributes *could* be required during the evaluation of a concrete access control request. Thus, the PEP resolves as much services attributes that are potentially required and sends them, proactively, together with the initial request to the PDP (see Fig. 1c).

PDP Prefetch: By deploying the pre-computed look-up-table only in the central PDP implementation, we avoid the need of customizing the various PEPs. As the PDP can easily check which service attributes might be necessary for a specific access control request, the PDP is able to request (in its first answer) all potentially required attributes at once. Therefore the number of evaluation attempts required is strictly bound to two and does not depend on the number of attributes required (see Fig. 1d).

2.3   Pre-computing Attribute Sets

In the following, we briefly discuss the steps required for pre-computing the mapping from access control requests to set of service attributes required by the two approaches based on pre-fetching. We assume that service attributes are not used within the core policy language, e. g., in case of XACML within the target match.

On an abstract level, a security policy P is a mapping of rules to access control decisions (e. g., deny or allow). For example, in case of RBAC with access control constraints, a rule is a four-tuple (g, r, a, c) where g is the required role, r is the resource, a is the action on that resource, and c is a constraint (i. e., a Boolean expression over context attributes). An access control request is a triple (u', r', a') where u' is the requesting user, r' is the resource the user is requesting access for executing action a'. Moreover, we say a rule (g, r, a, c) matches a request (u', r', a') if and only if, the user u' is a member of the role g, both $r = r'$ and $a = a'$ hold, and the access control constraint c' evaluates to true.

For PDP Prefetch, we need to compute a mapping from triple (g, r, a) to the set of potentially required service attributes. Given a policy P,

1. for each rule (g, r, a, c) in the policy, we compute a four-tuple $(g, r, a, A_s(c))$ where $A_s(c)$ is the set of service attributes that are syntactically referenced in the access control constraint c.
2. we group the set of four-tuples $(g, r, a, A_s(c))$ with respect to their lexico-graphic order of group, resource, and action (i. e., the triple (g, r, a)).
3. in each group from the previous step, we build die union of all required sets of service attributes, e. g., given $\{(g, r, a, A_s(c_1)), \ldots, (g, r, a, A_s(c_n))\}$, we compute the triple $(g, r, a, \bigcup_{1 \leq i \leq n} A_s(c_i))$.
4. we use the consolidated set of triples as input for initializing the hash-table used in the PDP implementation.

For the PEP Prefetch strategy, we compute analogously a mapping from resource-action tuples to the set of required service attributes.

While for policy written in an high-level language (e. g., SecureUML [4]) the set of attributes can be computed exactly, for complicated technical policies (e. g., XACML using various policy sets together with different policy combining algorithms) an over-approximation seems to be a good compromise avoiding the need for the semantical analysis of the given policy. For policy languages supported by a model-driven-security toolchain, e. g., [5], the required configurations (or even the complete PEP implementation) can be generated automatically.

3   Empirical Results

3.1   Scenario

We evaluate how the different strategies for resoling context attributes behave in three different scenarios, whereas these three scenarios differ in the size of the policies and the requirement to resolve service attributes:

Scenario I: This scenario uses a relatively small policy consisting out of 200 rules for ten resources, 50 users that are mapped onto fifteen roles, about 0.5 service attributes are resolved per access control request.

Scenario II: Compared to scenario I, this scenario has an increased likelihood that service attributes need to be resolved (about 1.8 per request), the policy has 500 rules for ten resources, 200 users are mapped onto 40 roles.

Scenario III: This scenario uses a large policy with 5000 rules for 100 resources, restricting the access of 800 users mapped to 100 roles. The policies extensively use service attributes, causing that an average access control request requires the resolution of over six service attributes.

For every scenario, we generated the policy, the user-group assignments, the required configurations, and a set of 2000 test requests. Attributes are resolved to a configurable random value. Generating the access control list explicitly allows for deterministic repetition of our benchmarks ensuring the comparability of the results for the different resolution strategies.

3.2   Experimental Results

We compare the different resolution strategies using a distributed prototype based on the Sun's XACML implementation (http://sunxacml.sourceforge.net) running on standard hardware. The PDP offers a Web service interface for the client PEPs and we use the AttributeFinderModule of XACML for implementing the PIPs used by the PDP.

First, for each attribute resolution strategy we compare the average response times for each scenario (see Fig. 2a). In general, this time reflects a significant portion of the time a system needs for reacting on a users actions. Second, we analyze the average network load, i. e., the amount of data transmitted during an access control request (see Fig. 2b).

For a small scenario (i. e., scenario I), we do not observe a significant difference in both the average response time and the average network traffic (see Fig. 2). For larger policies (i. e., scenario II), PDP Prefetch and PIP require twice as long as PEP Prefetch and the trial and error strategy takes more than four times longer. A similar behavior can be observed in scenario III.

Overall, PEP Prefetch is the fastest approach in all scenarios. Interestingly, the PIP is in scenario three faster than PDP Prefetch, although the PIP has to execute on average six Web service calls back to the client, compared to one additional XACML request done by PDP Prefetch.

The average network traffic shows a similar behavior: PEP Prefetch results in the minimal amount of network traffic per access control request. Both the trail and error strategy and the PIP are resulting in an increased network traffic that does grow larger than linear with respect to the increase of the number service attribute that need to be resolved. Thus, the overhead caused by Web service calls seems to exceed the overhead of sending additional attributes.

Further, we compare the average response time depending on the number of (required) service attributes. Fig. 3 shows that the overhead caused by PEP

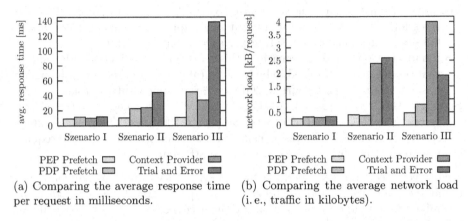

(a) Comparing the average response time per request in milliseconds.

(b) Comparing the average network load (i. e., traffic in kilobytes).

Fig. 2. Comparing the average response times and the overall network traffic for the different usage scenarios and resolution strategies

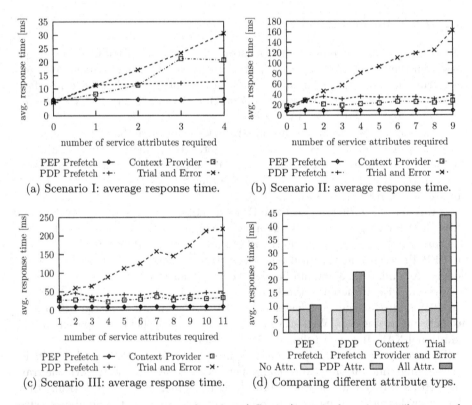

(a) Scenario I: average response time.

(b) Scenario II: average response time.

(c) Scenario III: average response time.

(d) Comparing different attribute typs.

Fig. 3. Comparing response times based on (effective) required service attributes, and comparing scenarios without attributes

Prefetch causes only minor delay in contrast to the three other approaches. In scenario I (Fig. 3a), PDP Prefetch results in the worst response time for one resolved service attribute. Albeit, it remains on this level for two and more service attributes. In contrast, the trial and error strategy has a similar response time as the PDP Prefetch approach for one service attribute, but the response time increases nearly linear with additionally required attributes.

For larger policies, i. e., scenario II and III, the size of the policy seems not to have a significant effect on how the different approaches behave relatively to each other (see Fig. 3b and Fig. 3c). Here, the PEP Prefetch strategy enjoys the fastest response time, which, moreover, also does not depend on the number of attributes required during the request evaluation.

Marshaling and unmarshaling the XACML (i. e., XML) seems to be a time consuming task: first, compared to Web service requests the effort for handling XACML seems to be very high. While for both the PDP Prefetch and the trail and error approach the same number of Web service requests are executed, the trial and error approach has a significantly slower response time.

Fig. 3d compares variants of scenario II using policies where all attributes are resolved on the PDP (i. e., causing no network overhead and delay), and without any attributes (except group resolution). As expected, using no service attributes is significantly faster. The four approaches differ only in service resolution significantly. The resolution of evaluation of local attributes (i. e., within the PDP) results in a minor overhead.

4   Discussion

Overall, our proposed resolution strategies (i. e., PEP Prefetch and PDP Prefetch) do not eliminate or replace the concept of PIPs. For resolving global or PDP attributes, a context provider directly attached to the PDP should be used.

The PEP Prefetch strategy for resolving context attributes from service side, produces only little overhead. Thus, even in scenarios rarely using service attributes, our approach is the fastest implementation—leading to the best user experience, i. e., fast response time of the system. The main drawback of the PEP Prefetch strategy is the fact that it requires an update of the PEP configuration whenever the security policy (i. e., the set of rules) requires additional service attributes for evaluation. The PDP Prefetch strategy overcomes this drawback while being only resulting in slightly higher response times and network traffic. As our implementation allows to use the PEP Prefetch strategy with a fallback strategy (PDP Prefetch or trial and error, e. g., for the time the updates need to be distributed to all PEPs after a policy update), we propose to use the PEP Prefetch strategy with a fallback strategy as a general solution.

Overall, the overhead for marshaling and unmarshaling XACML requests encoded in XML seems to be very high. Unfortunately, our experiments cannot give hints if this is an implementation specific or a general problem. Thus, a more efficient implementation may increase the overall performance in general and the performance of the PDP Prefetch approach in particular.

We assume that resolving attributes is possible with low costs. We think that this is a reasonable assumption (i. e., many attributes required will be assigned to or available within the current context of the access control request, e. g., owner of the accessed resource as attribute or property of the resource itself). Nonetheless, this cannot be guaranteed for all attributes. Especially for the context provider approach, this assumption is problematic: as the call back to the service is asynchronous, the resolution is not executed within the context of the access control request. Thus, the performance and costs of the resolution depends mainly on the underlying application: an intelligent session handling may allow a fast access of the request context and, therefore, allow the resolution with nearly the same costs as from within the context itself.

Classifying the attribute types depending in which parts of the overall system landscape they can be resolved efficiently seems to be a valuable add-on to existing policy frameworks. Overall, such a classification avoids, first, the need to transport any kind of status from the client over PDP and PIP to the attribute resolving component of the client. Second, in the context of the access control request they are for usual accessible without or with low costs.

5   Conclusion and Future Work

Work on improving the performance of PDPs in distributed service-oriented environments mainly focuses on three aspects: first, providing highly efficient PDP implementations for static policies (e. g., [13]), second, optimizing the evaluation performance by optimizing, statically, the policy (e. g., [14]). Third, while there is a large body of literature, e. g., [8, 10, 11], on improving the performance of security frameworks using caching strategies, only proactive caching [11, 12] addresses the caching of dynamic access control properties. Still, proactive caching is only able to cache dynamic access control constraints if they are first-class citizens of the underlying access control language.

As pre-computing the set of required attributes significantly helps in reducing both the delays caused by resolving context attributes and the overall network traffic, combining access control caching strategies with the attribute resolution approaches presented in this paper seems to be an attractive option for improving the overall performance of today's enterprise systems.

Furthermore, the performance overhead caused by encoding requests in XACML needs to be explored: either by switching to a different XACML policy decision engine, e. g., [13] or by looking at different policy decision languages and evaluation techniques.

Acknowledgments. This work has been supported by the German "Federal Ministry of Education and Research" in the context of the projects "SoKNOS" and "Polytos." The authors are responsible for the content of this publication.

References

[1] Anderson, A.H.: A comparison of two privacy policy languages: EPAL and XACML. In: ACM workshop on Secure Web services (SWS), pp. 53–60. ACM Press, New York (2006)

[2] Ashley, P., Hada, S., Karjoth, G., Powers, C., Schunter, M.: Enterprise privacy authorization language (EPAL 1.2). Tech. rep., IBM (2003), http://www.zurich.ibm.com/security/enterprise-privacy/epal

[3] Basel Committee on Banking Supervision: Basel II: International convergence of capital measurement and capital standards. Tech. rep., Bank for International Settlements, Basel, Switzerland (2004), http://www.bis.org/publ/bcbsca.htm

[4] Basin, D.A., Doser, J., Lodderstedt, T.: Model driven security: From UML models to access control infrastructures. ACM Transactions on Software Engineering and Methodology 15(1), 39–91 (2006)

[5] Brucker, A.D., Doser, J., Wolff, B.: An MDA framework supporting OCL. Electronic Communications of the EASST 5 (2006)

[6] Chadwick, D., Zhao, G., Otenko, S., Laborde, R., Su, L., Nguyen, T.A.: PERMIS: a modular authorization infrastructure. Concurrency and Computation: Practice & Experience 20(11), 1341–1357 (2008)

[7] Chen, H., Li, N.: Constraint generation for separation of duty. In: ACM symposium on access control models and technologies (SACMAT), pp. 130–138. ACM Press, New York (2006)

[8] Crampton, J., Leung, W., Beznosov, K.: The secondary and approximate authorization model and its application to Bell-LaPadula policies. In: ACM symposium on access control models and technologies (SACMAT), pp. 111–120. ACM Press, New York (2006)

[9] Kapsalis, V., Hadellis, L., Karelis, D., Koubias, S.: A dynamic context-aware access control architecture for e-services. Computers & Security 25(7), 507–521 (2006)

[10] Karjoth, G.: Access control with IBM Tivoli access manager. ACM Transactions on Information and System Security 6(2), 232–257 (2003)

[11] Kohler, M., Brucker, A.D., Schaad, A.: ProActive Caching: Generating caching heuristics for business process environments. In: Conference on Computational Science and Engineering (CSE), vol. 3, pp. 207–304. IEEE Computer Society, Los Alamitos (2009)

[12] Kohler, M., Schaad, A.: Pro active access control for business process-driven environments. In: Annual Computer Security Applications Conference (ACSAC) (2008)

[13] Liu, A.X., Chen, F., Hwang, J., Xie, T.: XEngine: A fast and scalable XACML policy evaluation engine. In: Conference on Measurement and Modeling of Computer Systems, Sigmetrics (2008)

[14] Miseldine, P.L.: Automated XACML policy reconfiguration for evaluation optimisation. In: Software engineering for secure systems (SESS), pp. 1–8. ACM Press, New York (2008)

[15] OASIS: eXtensible Access Control Markup Language (XACML) 2.0 (2005), http://docs.oasis-open.org/xacml/2.0/XACML-2.0-OS-NORMATIVE.zip

[16] Sarbanes, P., Oxley, G., et al.: Sarbanes-Oxley Act of 2002. 107th Congress Report, House of Representatives, pp. 107–610 (2002)

[17] Schaad, A., Spadone, P., Weichsel, H.: A case study of separation of duty properties in the context of the Austrian "eLaw" process. In: ACM symposium on applied computing (SAC), pp. 1328–1332. ACM Press, New York (2005)

Formal Verification of Application-Specific Security Properties in a Model-Driven Approach

Nina Moebius, Kurt Stenzel, and Wolfgang Reif

Institute for Software and Systems Engineering
University of Augsburg
86135 Augsburg, Germany
{moebius,stenzel,reif}@informatik.uni-augsburg.de

Abstract. We present a verification method that allows to prove security for security-critical systems based on cryptographic protocols. Designing cryptographic protocols is very difficult and error-prone and most tool-based verification approaches only consider standard security properties such as secrecy or authenticity. In our opinion, application-specific security properties give better guarantees. In this paper we illustrate how to verify properties that are relevant for e-commerce applications, e.g. 'The provider of a copying service does not lose money'. This yields a more complex security property that is proven using interactive verification. The verification of this kind of application-specific property is part of the SecureMDD approach which provides a method to model a security-critical application with UML and automatically generates executable code as well as a formal specification for interactive verification from the UML models.

1 Introduction

Verification of applications that are based on cryptographic protocols gives a better confidence in the security of these systems. Many approaches [PM08] exist that deal with the verification of security for generic cryptographic protocols, i.e. protocols that are not application-specific and can be used in different contexts (e.g. key exchange protocols). In contrast to that only a few approaches [BMP06] [HGRS07] exist that work on the security of cryptographic protocols which are tailored to a specific application. One example for such an application is a copycard system based on smart cards which is e.g. used at universities to provide a copy service. To verify the security of this application it is essential to formulate application-specific security properties like 'The provider of the copycard system does not lose money'. The property implies that no one can e.g. use a forged copycard to pay at the copying machines. This kind of security is usually needed if e-commerce applications that deal with electronic goods are considered.

In this paper we introduce a method to verify application-specific security properties using interactive verification and the theorem prover KIV [BRS+00].

F. Massacci, D. Wallach, and N. Zannone (Eds.): ESSoS 2010, LNCS 5965, pp. 166–181, 2010.

The verification approach is part of the project SecureMDD which presents a model-driven software engineering method to develop security-critical applications based on cryptographic protocols. Here, an application is modeled with UML (extended by a UML profile and a language called MEL (Model Extension Language) to describe the protocols on an implementation-independent level). Then, the models are used to automatically generate executable Java and Java-Card (for the smart card part) code. At the same time a formal specification is automatically generated from the UML models which can be imported by the KIV tool. An overview of the SecureMDD approach can be found in [MSGR09]. We illustrate our approach using the copycard as a small example and present a technique to cope with the verification. This work improves previous work done by our group [HGRS07].

The *generation* of the formal specification from a UML model has several benefits, and offers many possibilities. The application-dependent data types defined in the UML models are used directly, instead of a complicated generic data type. The formal specification is tailored to the application. For example, if the application does not use encryption this results in a formal specification which omits the parts relevant for encryption as well and thus simplifies the verification. Moreover, it becomes possible to check some (standard) properties automatically using a separate tool (e.g. a model checker), and use them as axioms in the proof of the main security property. All in all, the verification is now simpler, and much less time-consuming than before. The experiences and improvements are described in more detail in section 5.3.

The paper is organized as follows: Section 2 gives an overview of the copycard example and illustrates its modeling with UML. In section 3 the formal specification is described. Section 4 introduces the application-specific security property for the copycard application and Section 5 presents the verification technique. Section 6 discusses related work and Section 7 concludes.

2 Example: A Copycard Application

We illustrate the definition of application-specific security properties and their verification with an example. Our case study is a copycard application which can be used at universities to pay for copies at a copying machine. Each student owns a smart card, called copycard, onto which he can deposit money. To make copies, this copycard is inserted into a terminal (with a smart card reader) that is connected to a copying machine. Copies are paid by debiting their price from the card.

In the following the application and especially the security protocols to load money onto a copycard and to make copies are introduced. This is done by describing some parts of the platform-independent UML diagrams of the application. First, we describe the static view of the copycard system. Afterwards, the dynamic aspects, i.e. the security protocols, are explained. More information about the modeling of security-critical applications with UML, the used UML-Profile and the Model Extension Language (MEL) can be found in [MSR08]. One specialty of smart cards is that they are tamper-proof (which is a prerequisite for the protocols introduced here).

Figure 1 shows the class diagram of the copycard application. Central parts of the diagram are the classes *Terminal* and *Copycard* that represent the terminals to load money onto a card and to make copies as well as the copycard component. Note that we model only one kind of terminal that is able to debit as well as to deposit money. Both components have a *state* to store which message they expect to receive next (resp. how far they moved on through the protocol). The possible states are EXPLOAD (expecting a *Load* message), IDLE (expecting no specific message; after finishing a protocol run), EXPRESPAY (expecting a *ResPay* message) and EXPRESAUTH (expecting a *ResAuthenticate* message). Moreover, the copycard component has an attribute *balance* which stores the current amount of the card. Both, the terminal and the card, have an attribute *passphrase* that stores a *Secret* which is the same for all authentic components. Furthermore, both components have an attribute *challenge* that stores randomly generated nonces (to prevent replay attacks). Moreover, the terminal has a field *value* to store the amount which is loaded or debited in the current protocol run. The fields *issued* and *collected* are only used to express the security property and are explained in Chapter 4. The message formats used in the protocols are also defined in the class diagram. All subclasses of class *Message* describe one of the message types used in the protocol runs.

In Fig. 2 the communication channels are shown. A *Terminal* is able to exchange messages with a *Copycard* (while it is in the smart card reader which is connected to the terminal). Furthermore, the terminal receives messages from

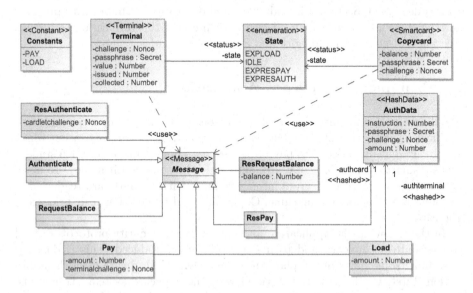

Fig. 1. The platform-independent class diagram of the copycard application

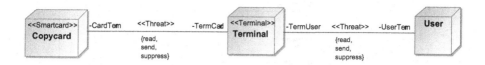

Fig. 2. Deployment diagram showing the communication infrastructure and attacker abilities

the *User*, who initiates a protocol run e.g. via a graphical user interface. The names associated with the association ends are called ports. These are used to distinguish the channels. If the attacker has access to a communication channel, the channel is associated with stereotype ≪Threat≫ that has three tags to indicate if the attacker can *read* messages that are sent over the channel, *send* messages or *suppress* messages. Then, modifying a message is also possible by removing a message from the channel and sending a similar one. In our example, we assume to have a Dolev-Yao attacker [DY81] who has full access to both channels.

The protocol to load money onto a copycard is shown in Fig. 3. The user initiates the protocol by inserting money (of amount *value*) into the terminal. In a next step, the terminal has to authenticate against the card. This prevents loading of money with a forged terminal and is done by a challenge-response authentication. In detail, the terminal checks whether the *value* is greater or equal zero, sets its *state* to *EXPRESAUTH* and sends an *Authenticate* message to the card to request a challenge. The card updates its *state* to *EXPLOAD* and sends back a newly generated nonce (stored in field *challenge*). The terminal checks and resets its state and generates a hash value over a constant (that ensures that the hash value belongs to a Load protocol run), the *passphrase* (the secret which is known to all authentic components), the received challenge and the *value* to be loaded. Then, the terminal sends a *Load* message, containing the amount to load (= *value*) and the hash value to the card. The card checks its state as well as the hash value. If both are ok, the value is added to the balance of the card.

Fig. 4 shows the protocol to make copies. The user tells the terminal the number of copies he wants to make (resp. the corresponding amount he has to pay). In this protocol, the card has to authenticate against the terminal to ensure that no forged card is used to make copies. Again, we use challenge-response authentication. The terminal sets its *state* to *EXPRESPAY* and sends a *Pay* message (containing the amount and a newly generated nonce) to the copycard. The smart card reduces its balance and generates a hash value in the same way as the terminal in the Load protocol. Then, the hash value is sent back to the terminal. If the terminal is in state EXPRESPAY and the received hash value is ok, the payment is accepted and the copying machine makes the copies.

Furthermore, there exists one protocol to request the current balance of a card. This protocol is very simple and omitted here.

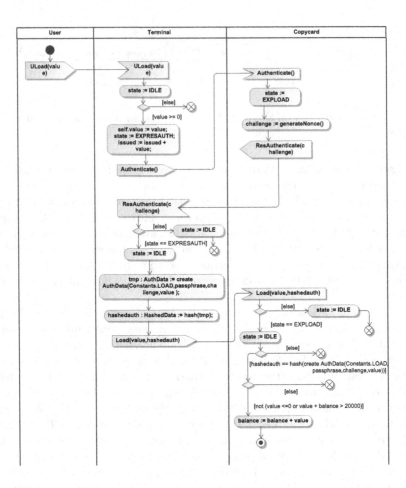

Fig. 3. Activity diagram showing the Load Protocol

3   Generating a Formal Specification

In this section we describe a part of the formal specification of the copycard application which is generated from the UML models and directly used as input for our interactive theorem prover KIV. We present an overview of the part of the specification which is needed to understand section 5. More detailed information about the generation of the formal specification from UML can be found in [MSR09].

In contrast to other approaches our formal specification covers the whole application. The static part of an application consists of the components of the system, the used data and message types, the communication infrastructure as well as the abilities of the attacker. This information is contained in the class and deployment diagrams in UML. In the formal model the static parts are defined using algebraic specifications. The components of the copycard application are

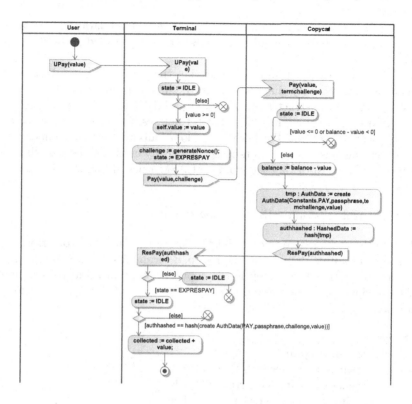

Fig. 4. Activity diagram showing the Protocol to make copies

the copycards, the terminals, an attacker (that attempts to interfere with the protocol runs) as well as a user that initiates the protocol executions. In the formal model the component types smart card and terminal have an arbitrary but finite number of instances. This models the fact that several terminals and copycards exist in real world applications. The data and message types defined in the class diagram are used in the formal specification as well. Additionally, data types for encrypted, hashed and signed data as well as cryptographic operations such as encrypt, hash and so on are generated. Since only some components are able to communicate, we explicitly model the existing communication channels. Each channel has two endpoints consisting of the component as well as a port that is used to distinguish the communication channels. Each endpoint has an inbox which is given as a list of messages. The inbox contains the messages that were sent to a component but are not yet processed. In the formal specification the sending of a message over a channel is equal to adding it to the inbox of the corresponding endpoint. An important part of the formal model is the specification of the attacker. The attacker is modeled as a component which is associated with a set of data, his knowledge. If a message is sent over a channel that the attacker is able to read, this message is added to his knowledge and it is analyzed whether the attacker knowledge has to be extended. E.g. if the message contains a key and the attacker already had an encrypted data in his knowledge that can be decrypted by using this key now, he gains knowledge about the plain data.

The dynamic part of the application, i.e. the security protocols, are given as sequence and activity diagrams in UML. In the formal model this part is translated into an abstract state machine (ASM)[BS03]. The main ASM rule called ASM consists of a while loop that terminates if a stop flag is set to true. Initially this flag is false. The body of the loop calls an ASM rule named STEP. After executing this rule it is non-deterministically chosen if the stop flag is set to true or false. Note that the names of ASM rules are written in uppercase letters.

ASM { while ¬ stop do {STEP; choose stop;}}

The ASM rule STEP non-deterministically chooses an action to perform. Possible steps are a user-step, an attacker-step, a terminal-step or a copycard-step. For example, if the chosen step is the copycard-step, an arbitrary component c of type copycard is chosen and the ASM rule COPYCARD is called for this component.

```
STEP {
    choose asm-step in {
        if asm-step = user-step then
            {choose c with isUser(c) in USER}
        else if asm-step = attacker-step then
            {ATTACKER}
        else if asm-step = terminal-step then
            {choose c with isTerminal(c) in TERMINAL}
        else if asm-step = copycard-step then
            {choose c with isCopycard(c) in COPYCARD}
        else ...
    };};
```

The COPYCARD rule stores the first message in field inmsg and removes it from the inbox. If the inbox is empty, the COPYCARD rule terminates. Then, the type of the message inmsg is checked and depending on the result the ASM rule to process this message is executed. If inmsg is a *Load* message the rule LOAD is called. This rule processes the message and updates the state of the copycard component, i.e. sets its state to IDLE and increases the balance of the card. One step of the ASM is equivalent to one step of the protocol. In the UML models the protocol steps can be found in the activity diagrams. Each step starts with the receiving of a message, followed by some checks, updates and finally the sending of a message or an activity final node in case the end of the protocol is reached.

```
COPYCARD {
  let inmsg = inputs(c)(inport) .first in {
      inputs(c)(inport) := inputs(c)(inport).rest;
      if isRequestBalance(inmsg) then REQUESTBALANCE
      else if isLoad(inmsg) then LOAD
      else ...
      else ABORT }
};
```

In case the attacker was chosen instead of a copycard component the ASM rule for the attacker is called. Then, it is non-deterministically chosen if the attacker suppresses or sends a message. If the attacker suppresses a message a non-empty inbox is chosen that belongs to a channel where messages can be suppressed by the attacker. Then, several randomly chosen messages are deleted from that inbox. In the send case the attackers generates an arbitrary message from his current knowledge. The message is then sent to an (randomly chosen) inbox reachable by the attacker (i.e. the channel has the attacker-send property). For example, if the attacker knows a secret and a nonce, he can use them to generate an object of type *AuthData*, hash this object and send a *Load* message containing this hash value. Or, if the attacker already knows a hash value, he can use it to send a *Load* message. Note that the attacker is not able to generate arbitrary secrets and nonces that are not already stored in his knowledge. The specification of the attacker is similar to the attacker in the inductive approach of Paulson [Pau98].

4   Security Properties

The security properties we are interested in are business-specific and depend on the application. Most existing tool-based verification approaches (e.g. [Jan05], [Mea96]) that deal with verification of cryptographic protocols consider and prove standard security properties such as secrecy, integrity or authentication. In our opinion, this does not suffice to ensure the security of protocols in certain applications. For example, in case of the copycard application, the operator of the copying machines is interested in a verified statement that says 'It is impossible to spend more money on the copying machine than was loaded onto a card by using a real terminal (and inserting real money into this terminal).' If this holds, the operator knows that the user of the copying service is not able to defraud and e.g. use a forged terminal to load money onto his card or change the balance of the card. This kind of security property always depends on the application. Although we do not focus on proving standard security properties, some of them are usually needed as preconditions for the main security proof.

When designing a security protocol one always has to keep in mind that there are several participants with different security concerns. In case of the copycard, we focused on the security property which is important for the operator of the copying machines. For example, the Load protocol (introduced in Fig. 2) does not ensure that the amount of money which was inserted into the Load terminal is really loaded onto the card. An attacker could suppress the *Load* message which is sent from the terminal to the copycard. In this case, the owner of the card would not get back his money and would not even notice that the loading of money was not successful.

To express that the attacker is not able to defraud we add two attributes to the terminal class. One attribute named *issued* which counts the amount of money that was inserted into the terminal by a user. This field is increased after the terminal received a *ULoad* message from the user. This message models the inserting of money and has the inserted amount as argument (= *value* and *issued* is increased by *value*). Each terminal has a field *issued* and we are interested in the sum of money that was inserted into all terminals. Assuming that numterm is the number of existing terminals we define the axiom allIssued as follows: $\text{allIssued}(\text{issued}) = \sum_{n=1}^{numterm} \text{issued}(\text{terminal}(n))$. In contrast to that we count the amount of money that was spent by all users to make copies. Thus, we also define an attribute named *collected* that increases at the end of the protocol to make copies (after receiving a *ResPay* message from the card and verifying the hash value). Again, we consider all terminals and compute the sum of money which was collected by all terminals connected to a copying machine: $\text{allCollected}(\text{collected}) = \sum_{n=1}^{numterm} \text{collected}(\text{terminal}(n))$.

Remember that our aim is to verify that the user of the copying service is not able to defraud. This can be expressed as 'The sum of money that was inserted into the load terminals (and issued to the cards) is always greater or equal than the amount that was spent to make copies (and collected by the terminals)': allIssued $\geq$ allCollected.

5   Verification Technique

5.1   Verification of the main Security Property

Our goal is to prove that the main security property allIssued(issued) $\geq$ allCollected(collected) always holds. Thus, we have to prove that after executing any possible sequence of protocol steps the security property still holds:

$$\text{init}(..) \tag{1}$$
$$\vdash [\text{ASM}(..)] \ \text{allIssued(issued)} \geq \text{allCollected(collected)}$$

$[\cdot]$ is the box operator of dynamic logic. The meaning of $[\alpha]\varphi$ is that if program α terminates the condition φ holds afterwards. We start with an initial state (i.e. balance is set to zero, the state of all terminals and cards is IDLE, the attacker-knowledge is empty and so on). Then, the protocol steps are chosen non-deterministically and executed. Thus, we consider all finite sequences of steps.

To prove obligation (1) we first verify that the security property is invariant with respect to all possible steps of the application. We (try to) prove that, under the assumption that some basic invariants and the security property hold, every STEP (see section 3) terminates and the security property still holds afterwards:

$$\text{BASE-INV}(..) \wedge \text{allIssued}(\text{issued}) \geq \text{allCollected}(\text{collected}) \tag{2}$$
$$\vdash \langle\!\langle \text{STEP}(..) \,\rangle\!\rangle \; \text{allIssued}(\text{issued}) \geq \text{allCollected}(\text{collected})$$

$\langle\!\langle \cdot \,\rangle\!\rangle$ is the strong diamond operator of dynamic logic. The meaning of a formula $\langle\!\langle \alpha \rangle\!\rangle \; \varphi$ is that all runs of the program α terminate and the condition φ holds afterwards (this corresponds to $\text{wp}(\alpha,\varphi)$ in Dijkstra's wp-calculus). To be able to prove this invariant, we first have to prove that several other invariants hold. This includes lemmas dealing with nonces, e.g. the field *challenge* of all terminals and copycards never contains the same nonce and the balance of all cards is greater or equal to zero at any time. Other invariants state that all 'real' cards and terminals have the same *passphrase* and the attacker does not know it. All in all, there are 10 invariants that are grouped by the BASE-INV invariant.

Unfortunately, proof obligation (2) does not hold. The reason is that the inequality allIssued(issued) $\geq$ allCollected(collected) is not invariant for all single steps. For example, when a *ResPay* message is received and processed, the field *collected* is increased by the value of copies the user makes. The information that the amount to make copies is less or equal to the amount that was loaded (and added to *issued*) earlier is lost because we only consider single steps of the protocol.

The solution is to extend the security property and keep the knowledge of every change in the state of the terminals and copycards that has influence on the fields *issued* and *collected*. For example, if the field *issued* is increased by a value, the *balance* of a copycard is increased by the value as well (in the next step). If *collected* is increased, the *balance* of one card is decreased by the same value in the step before. Thus, we consider the balances of all cards as well. Assuming that `numcards` is the number of all copycards that are in use, we define: allBalance(balance) = $\sum_{n=1}^{numcards}$ balance(copycard(n)) and get the generalized security property allIssued(issued) $\geq$ allCollected(collected) + allBalance(balance).

In the Load protocol the terminal receives an *ULoad* message, updates its field *issued* and sends a *Load* message to the copycard. In another step, the copycard receives this message (if it is not suppressed by the attacker) and updates its *balance*. Since these updates do not happen in the same step, we have to consider the values of the valid *Load* messages that are 'in transit' (= are in an inbox of the copycard) as well because they may cause an update of the balance in a later step. Furthermore, it is necessary to consider the values of the valid *Load* messages that were suppressed by the attacker but could be replayed by him. Here, it is utilized that we have a Dolev-Yao attacker and all messages that are in an inbox of a copycard or terminal are also element of the attacker knowledge. Thus, we ignore the messages that are stored in the inboxes and only consider the valid *Load* messages known to the attacker. We have to determine the most 'valuable' *Load* message the attacker can generate for a given nonce and secret. The attacker is able to generate a *Load* message with *nonce*, *secret* and value i if he has the corresponding hash value in his knowledge. We use a predicate to determine whether the attacker knows at least one hash value:

attackerHasLoadHash(attacker-known, nonce, secret)
$\leftrightarrow (\exists\, i \geq 0.\ \text{hash}(\text{mkAuthData}(\text{LOAD},\text{secret},\text{nonce},i)) \in \text{attacker-known})$

During the protocol execution it may happen that two or more valid *Load* messages exist for one card. If the attacker has more than one valid hash value in his knowledge, the one with the highest value is considered. This is sufficient because this is the one that causes the highest increase of the field *balance*.

getLoadValue(attacker-known,nonce,secret) =
$\max\{i : \text{hash}(\text{mkAuthData}(\text{LOAD},\text{secret},\text{nonce},i)) \in \text{attacker-known}\}$

The maximum of the empty set is zero. All *Load* messages that are known to the attacker and which are currently accepted by a copycard have to be considered. A copycard accepts a *Load* message if its state is set to EXPLOAD (= expecting Load message) and the hash value of the message (over the copycards current challenge and passphrase) is correct (and in the knowledge of the attacker). We have to sum up the values of all *Load* messages that are accepted by any card. This is done using a recursive function called validLoadMessages. Remember that we have `numcards` cards (with index n from 1..`numcards`).

attackerHasLoadHash(attacker-known,
                 challenge(cardlet(n)), passphrase(cardlet(n)))
$\wedge$ state(cardlet(n)) = EXPLOAD
$\rightarrow$ validLoadMessages(state,challenge,passphrase,attacker-known,n) =
     getLoadValue(attacker-known,
                 challenge(cardlet(n)), passphrase(cardlet(n)))
   + validLoadMessages(state,challenge,passphrase,attacker-known,n-1)

validLoadMessages(state,challenge,passphrase,attacker-known,n) sums up the values of all *Load* messages accepted by the copycards 1..n. n is the index of the copycard, in the beginning n is set to `numcards`. If the attacker has a *Load* message that is accepted by copycard(n) and this card is in state EXPLOAD, the sum that counts the valid load values is increased by getLoadValue (which returns the highest positive value of all valid *Load* messages for card n) and the function validLoadMessages is computed for the remaining n-1 copycards.

If the considered copycard is not in state EXPLOAD or the attacker has no valid *Load* message for this card, the sum remains unchanged for this card and the validLoadMessages are computed for the remaining cards 1,..,n-1.

In the same way as in the Load protocol we have to consider the value of the *ResPay* messages. Similar to Load, the *balance* of the card is decreased in the Pay step while the field *collected* of the terminal is increased in step ResPay. After executing the Pay step and before processing the corresponding *ResPay* message this message is stored in the inbox of the terminal or was suppressed by the attacker. Analogue to the function validLoadMessages we define a function validResPayMessages that considers the *ResPay* messages which are currently accepted by an existing terminal and computes the sum of all values that are added to the fields collected if the valid *ResPay* message is sent to the corresponding terminal.

A last complication is that we have to count the values of the terminals that were already added to the field *issued* but that are not yet stored in the valid-LoadMessages because we first have to do the authentication with the copycard (Authenticate and ResAuthenticate steps). Again, we consider all existing terminals and if a terminal is in state EXPRESAUTH we count its value.

state(terminal(n)) = EXPRESAUTH

$\rightarrow$

allValues(..,n) = value(terminal(n)) + allValues(..,n-1)

If terminal(n) is not in state EXPRESAUTH, the value of this terminal is ignored while computing allValues. Our modified security property looks as follows:

modSecProp:
    allIssued(issued)   (3)

$\geq$

      allCollected(collected)
      + allBalance(balance)
      + validLoadMessages(..,attacker-known,numcards)
      + validResPayMessages(..,value,attacker-known,numterms)
      + allValues(value,state,numterms)

Using this inequality all relevant changes of fields that may result in a modification of the fields *issued* and *collected* in later steps are considered. For example, if the field *issued* of one terminal is increased, the validLoadMessages are increased by the same value in the same step. If the balance is reduced by value v, the validResPayMessages are increased by the same value in the same step. If the corresponding *ResPay* message is processed, the validResPayMessages are decreased by v (because the terminal sets its state to IDLE) and the field *collected* of this terminal is increased by v in the same step.

Note that the equality of the equation does not hold because the attacker is able to suppress a valid *Load* message and e.g. send an *Authenticate* message to the same card. In this case, the *Load* message is invalid, is not counted by validLoadMessages anymore and cannot be used in the future.

Now, we are able to prove that the modified security property is invariant (see (4)) using the symbolic execution strategy for KIV dynamic logic.

BASE-INV(..) $\wedge$ *modSecProp* $\vdash$ ⟨|STEP(..) |⟩ *modSecProp*   (4)

Then, it is easy to prove formula (1) with an invariant rule.

5.2 Auxiliary Properties

To prove the main security property several lemmas are required. These can be grouped in four categories:

- Lemmas that are valid for all applications e.g. those only depending on the model of the attacker, communication and so on. Examples are that the function generateNonce() never returns a nonce that was already generated and that data which can be generated from the attacker knowledge can still be generated if something is added to the knowledge. These theorems are part of a library which is available for all applications.

- Lemmas that are generated from the UML models. They depend on the application but are obviously true and no proof is needed. One example is that if the attacker is able to read messages sent over any channel of the application, he knows all messages that are in an inbox of a terminal or copycard.

- Application-dependent standard security properties like the attacker never gets any knowledge about the common secret of all terminals and copycards. These properties can be proved using interactive verification but maybe one could also use model-checking to automatically prove these properties. This is future work.

- Application-dependent lemmas that cannot be reused for other applications. An example for the copycard is that the balance of each card is always greater or equal zero.

5.3   Experiences

The verification of the copycard application took several weeks. This includes the time that was needed to find design errors in the protocols and to rebuild the proofs for the modified protocols.

One difficulty while doing the verification was to find the appropriate invariants. For example, it is not obvious which properties are needed as basic invariants for the security proofs. The fact that the balance on the card is never negative is needed, but the fact that the value contained in a valid load message also is never negative is *not* needed. Furthermore, finding the invariant for the security proof (with valid Load messages and considering the values) was difficult and took several iterations. This is mostly due to the fact that no equations hold between the different values, but only inequalities since the attacker can destroy money.

The number of interactions needed for the proofs is about 3000 in total, about 2500 for the security proofs and about 500 for the lemmas. This is about one third of the effort needed previously (in [HGRS07]), and can still be improved.

More important than improvements in the sheer numbers are, however, the improvements in the simplicity and understandability of the formal model, and the seamless tool support. The complete formal specification is now generated automatically from UML models that are tailored to the design of cryptographic protocols [MSR08]. The models are easy to read and understand. For the copycard application, we generated 44 specifications, and the ASM is about 350 lines long. However, since there is a simple mapping between the description of the protocol steps (with activity diagrams and MEL) and the ASM steps, it is not necessary to look at the ASM rules at all. It is much more convenient to look at the UML models while doing the verification. If verification fails and an error is detected in the protocol it is corrected in the UML model, and the formal model is updated automatically. The proofs done so far are preserved if they are still valid (this is checked by the correctness management of the KIV system).

Another major improvement was the shift from a generic data type called Document to application-specific data types defined in the UML models. The Document type is recursively nested to allow its reuse for all kinds of protocols and messages. This requires a huge library of lemmas that are very tricky to prove (e.g. for the computation of the current attacker-knowledge), and makes the formulas very difficult to read. The application specific data types are usually not recursive, and not very deeply nested. Most operations become trivial, and the few lemmas that are needed are generated automatically.

Another improvement is that the generated formal specification is tailored to the application. For example, if the application does not use encryption, the computation of the attacker-knowledge is much easier. The communication structure is fixed based on the deployment diagram (see Fig. 2).

All this is only possible because of the model-driven approach based on model transformations and generation of the formal specification. It is part of our overall approach to develop security-critical applications called SecureMDD. In addition to the verification this also includes the generation of the full, runnable code for the terminal as well as Java Card code which is executable on a real smart card. The transformations that generate the formal specification are implemented using the model-to-text language XPand (which is part of the Eclipse Modeling tool).

6   Related Work

A lot of verification techniques and tools to prove the security of cryptographic protocols exist, a recent overview is given in [PM08]. These techniques can be divided into three categories: belief logics (e.g. [BAN90]), state exploration (e.g. [BMV05],[Low96]) and theorem proving (e.g. [Pau98], [Mea96], [Bla09]). Most of them are based on automatic tools and focus on generic security protocols (which are not specific to an application, e.g. authentication protocols) and prove standard security properties. In contrast to that the protocols of the copycard system as well as the security property highly depend on the considered application. It is not clear if automatic tools can cope with that kind of security property.

Some approaches dealing with application-dependent security properties exist. One method that is related to ours is the inductive approach of Paulson [Pau98] that uses the theorem prover Isabelle for verification and was successfully applied to several case studies. Bella extends the inductive approach to deal with smart cards [Bel01] but concentrates on generic security protocols as well. In [BMP06] Bella, Massacci and Paulson give an overview of their work on SET (Secure Electronic Transaction), a set of e-commerce protocols devised by Visa and Mastercard. The case study is formally modeled and verified using the inductive approach. Considered and proven (application-specific) security properties are that the payment information of the customer are only known to the bank, not to the merchant, and that the order information is not known to the bank.

Jürjens [Jan05] developed a method to model systems based on cryptographic protocols with UML. Inputs for model checking and automatic theorem proving are automatically generated from the models and standard security properties are proven. Besides several other case studies Jürjens worked on the security of the Common Electronic Purse Specification (CEPS) for cash-free point-of-sale transactions. One considered security property is that the sum of balances of all cards (that are used for payments) and all cards of the merchants (where the earned money is stored) is the same at any time. The appropriate proof was not done by tools but is paper-based [Jür04].

Another interesting case study with application-specific security properties is the Mondex Electronic Purse. Mondex has received a lot of attention because its formal verification has been set up as a challenge for verification tools [Woo06] that several groups worked on. The results of the participating groups are summarized in [JW08]. Mondex is about transferring money (from one card to another) but, except for our group, the participating groups only considered protocols without cryptography.

7   Conclusion

We presented a technique to interactively verify application-dependent security properties for applications based on cryptographic protocols, as an example we used a copycard application. Here, one interesting application-dependent property is that the provider of the copying service does not lose money. This kind of security property gives better confidence in the security of cryptographic protocols than standard properties, e.g. secrecy and integrity. The verification is part of the SecureMDD approach which is a model-driven method to develop security-critical applications. The formal specification used for interactive verification is automatically generated from the UML models that are used to specify the system under development.

References

[BAN90]   Burrows, M., Abadi, M., Needham, R.: A logic of authentication. ACM Transactions on Computer Systems 8(1), 18–36 (1990)

[Bel01]   Bella, G.: Mechanising a Protocol for Smart Cards. In: Attali, S., Jensen, T. (eds.) E-smart 2001. LNCS, vol. 2140, p. 19. Springer, Heidelberg (2001)

[Bla09]   Blanchet, B.: Automatic Verification of Correspondences for Security Protocols. Journal of Computer Security 17(4), 363–434 (2009)

[BMP06]   Bella, G., Massacci, F., Paulson, L.C.: Verifying the SET purchase protocols. Journal of Automated Reasoning 36(1-2), 5–37 (2006)

[BMV05]   Basin, D.A., Mödersheim, S., Viganò, L.: OFMC: A symbolic model checker for security protocols. Int. J. Inf. Sec. 4(3), 181–208 (2005)

[BRS+00]   Balser, M., Reif, W., Schellhorn, G., Stenzel, K., Thums, A.: Formal system development with KIV. In: Maibaum, T. (ed.) FASE 2000. LNCS, vol. 1783, p. 363. Springer, Heidelberg (2000)

[BS03]   Börger, E., Stärk, R.F.: Abstract State Machines—A Method for High-Level System Design and Analysis. Springer, Heidelberg (2003)

[DY81]   Dolev, D., Yao, A.C.: On the security of public key protocols. In: Proc. 22th IEEE Symposium on Foundations of Computer Science. IEEE, Los Alamitos (1981)

[HGRS07]   Haneberg, D., Grandy, H., Reif, W., Schellhorn, G.: Verifying Smart Card Applications: An ASM Approach. In: Davies, J., Gibbons, J. (eds.) IFM 2007. LNCS, vol. 4591, pp. 313–332. Springer, Heidelberg (2007)

[Jan05]   Jürjens, J.: Secure Systems Development with UML. Springer, Heidelberg (2005)

[Jür04]   Jürjens, J.: Developing high-assurance secure systems with UML: a smartcard-based purchase protocol. In: IEEE International Symposium on High Assurance Systems Engineering (2004)

[JW08]   Jones, C., Woodcock, J. (eds.): Formal Aspects of Computing, vol. 20 (1). Springer, Heidelberg (2008)

[Low96]   Lowe, G.: Breaking and Fixing the Needham-Schroeder Public-Key Protocol Using FDR. In: Margaria, T., Steffen, B. (eds.) TACAS 1996. LNCS, vol. 1055, pp. 147–166. Springer, Heidelberg (1996)

[Mea96]   Meadows, C.: The NRL protocol analyzer: An overview. Journal of Logic Programming 26(2), 113–131 (1996)

[MSGR09]   Moebius, N., Stenzel, K., Grandy, H., Reif, W.: SecureMDD: A Model-Driven Development Method for Secure Smart Card Applications. In: Third International Workshop on Secure Software Engineering, SecSE, at ARES 2009. IEEE Press, Los Alamitos (2009)

[MSR08]   Moebius, N., Stenzel, K., Reif, W.: Modeling Security-Critical Applications with UML in the SecureMDD Approach. International Journal on Advances in Software 1(1) (2008)

[MSR09]   Moebius, N., Stenzel, K., Reif, W.: Generating Formal Specifications for Security-Critical Applications - A Model-Driven Approach. In: ICSE 2009 Workshop: International Workshop on Software Engineering for Secure Systems (SESS 2009). IEEE/ACM Digital Libary (2009)

[Pau98]   Paulson, L.C.: The Inductive Approach to Verifying Cryptographic Protocols. J. Computer Security 6 (1998)

[PM08]   Lopez Pimental, J.C., Monroy, R.: Formal support to security protocol development: A survey. Computacion y Sistemas 12(1) (2008)

[Woo06]   Woodcock, J.: First Steps in the Verified Software Grand Challenge. IEEE Computer 39(10), 57–64 (2006)

Idea: Enforcing Consumer-Specified Security Properties for Modular Software

Giacomo A. Galilei and Vincenzo Gervasi

Dipartimento di Informatica, Università di Pisa

Abstract. Nowadays systems that download updates from the net or let the user download third-party code for extending the application functions (plug-ins) are widespread. In these dynamic environments the code that is going to be executed is not known at compile-time, and often not even at application start-up, neither by the application producer nor by the user. This turns reliable, well designed software into a dangerous and potentially malicious software for the user and for the system it runs onto: i.e., a well-behaved modular application becomes the unwilling host for malicious components. In this scenario, the application producer lines up with the user in requesting that dynamically loaded third-party components must satisfy given security requirements.

In this paper we present a framework that allows the consumer side of untrusted code to state desired properties about it. We exploit the facilities of the so-called virtual execution environments to encode directly into the meta-data of object code a well structured specification. Once the dynamic component is loaded at run-time by the main application, the framework will recover such specifications and check them against the requirements gathered from the main application, the user and the host operating system, injecting run-time checks as needed into the untrusted code to ensure that the actual behaviour of the component matches the specified one.

1 Introduction

The spread of modular applications and component-based programming has rendered some well-known analysis approaches [1] ineffective for modern applications. Many large applications in widespread use (e.g., integrated development environments such as Eclipse or Microsoft Visual Studio; image processing software such as Gimp or Adobe Photoshop; web browsers such us Firefox or Microsoft Internet Explorer) derive a large chunk, if not most, of their usefulness from third-party components (plug-ins). For such applications, the code that is going to be executed is in general not known at application compilation time, and often (due to dynamic loading techniques), even at application startup time. To make things worse, plug-ins can be downloaded (even repeatedly, e.g. in the case of auto-updating software) off the net, and hence be totally unreliable.

This situation calls for improved security assurance techniques, where the application producer lines up with the user in requesting for dynamic code to satisfy security requirements.

F. Massacci, D. Wallach, and N. Zannone (Eds.): ESSoS 2010, LNCS 5965, pp. 182–191, 2010.

In this paper we introduce an idea to allow the consumer side of untrusted code (i.e., the user and the host application) to state desired properties about it. We present a solution based on the Java language (and JVM) and on relatively recent developments such as *Annotations* and *Reflection*, but the idea is well applicable to any other language that runs on the top of so-called *virtual execution environments* that provide the same kind of functionalities, like the CLR and the .NET languages.

Our approach is based on the application of two different techniques. On the producer side, the code can be annotated with its own specifications; these can refer to declared properties of the software (e.g., in a well-meaning plug-in) or to requested properties of a component (e.g., in the host application, whenever a plug-in is invoked). A well structured language for specifications like JML [2] can be used at this stage. The JML specifications are parsed at compile-time to create an *Abstract Syntax Tree* (AST). Through the *Annotations* we attach a serialized form of the AST directly onto the meta-data of the object; this makes the specification fully and readily recoverable at run-time.

On the consumer side instead, we introduce a wrapper layer, which loads the host application and, through *Reflection*, retrieves the application's requirements on the plug-ins, and injects appropriate code hooks to verify them (and others) at run-time. Static verification techniques can also be applied to reduce the number of run-time checks.

The paper is organized as follows. The next section provides a brief overview of the problem we are addressing, while Section 3 describes the technical contribution of our framework. An example scenario is provided in Section 4, followed by some conclusions and by an account of the current state of the work.

2   Security Properties for Modular Software

We address a scenario where a modular application A is executed on a system S (which includes the operating system and libraries as well as the actual machine), on behalf of a user U. The application itself is modular in that it relies on one or more plug-ins P to accomplish whatever U needs A for. The producers of A, S and P are in principle mutually distinct, and do not trust each other. U trusts S, may trust A, and typically cannot trust P.

The code that is actually executed is given by $A + P$ (the actual code paths will include portions of S, but since U trusts S, we can disregard them for our purposes). Also, S will typically enforce its own requirements on $A + P$ by employing typical security mechanisms at the OS level (e.g., isolated memory spaces for different processes, file-system level security, privileged and non-privileged execution modes etc.), so we can also disregard that part. Our problem is then how to make sure that the requirements of U on $A + P$ and of A on P are satisfied, respectively, by $A + P$ and P. This problem is not totally new; in fact, it bears some similarity with the problem of ensuring reliability of libraries used in traditional applications. However, there are several factors that make standard security approaches ineffective or impractical in this case:

1. Source-based trust policies have been attempted for specific devices (e.g., the Apple Store for iPhone applications - where we can consider the applications as plug-ins of the front-end running on the device), but they are effective only in situations where a monopoly can be accepted — which is not the case in most context. For example, similar applications for PocketPC or SmartPhone devices (which can be downloaded from arbitrary sources) are hard to verify, and require a significant leap of faith on the part of the owner of the device. Having a single source for plug-ins in modular applications almost defeat the purpose, which is leveraging the creativity of a large number of independent developers to provide a rich ecosystem of additional functions.

2. Cryptographic signatures have also been used to ensure that a package, once shipped by the producer, is not altered, and moreover that the identity of the producer can be established (both for trust purposes and to ensure accountability). This family of techniques also provides only a partial solution; on one side, they require a shared and pervasive Public Key Infrastructure system, which is not widespread in the current market. Second, they can only work if there is a trust relationship between the user and the producer, which is a reasonable assumption for large, well-known producers (e.g., Microsoft), but not for the many, small, and relatively unknown producers to which a plug-in architecture appeals.

3. Most current techniques assume that the security properties that will need to be verified are the same for all users, and can thus be established once and for all by the producer, acting as a proxy for his customers. But in practice, different users have different security requirements (see Section 4), and any trust relationship established on the basis of the identity of the producer or of the source for the executable code cannot cater for such cases.

On the other side, emerging technologies provide new opportunities for addressing our problem. Virtual execution environments with their rich, type-safe languages; language features such as annotations; standardized deployment formats including rich meta-data capabilities let us leverage new opportunities that were not available till a few years ago. In particular, we will use these capabilities to let the producer of A state his requirements on P in the form of meta-data embedded in A's code; to let P provide a (partial) specification of its own behaviour also as metadata, and to dynamically alter the executable code of A and P so that U's requirements can be checked for at runtime by appropriate instrumentation.

3   The Framework

Given the four actors of our framework, the plug-in P, the host application A, the application user U, and the system S on top of which the application runs, we can identify two groups. On one side the code producer is interested in making its product (the plug-in) trustable; on the other side the code consumers (the application, the user and the system) are interested in specify requirements that the plug-in must satisfy.

Our approach to the problem operates over these two fronts. The framework supplies facilities to the code producer that yield a tailored object file which includes its own specification, while on the code consumer side it will make available a wrapper for running the application in a safer environment. The wrapper, which provides an API to the application, will make it possible to intercept the plug-ins loading process, to inspect their code, to verify it and to grant or deny execution. If execution is granted, run time checks can be injected into P's code to ensure that required properties are satisfied.

There are two steps in plug-in verification: the first one is performed to make the plug-in specification trustable; the second one is performed to ensure that the module will behave in accordance to what required by U, A and S. Note that the first step will also be used to reduce the scope of the second step of verification, since we will not verify twice the same property. This means that the more deeply specifications are detailed, the less costly the second step of verification will be.

3.1   Code-Producer Side

At compilation-time our framework will let the producer specify the module behavior. We exploit the *JML* [2] language for this purpose as shown in Section 3.1. Such specifications will then be tailored into a Java object file to be carried to the JVM. We are interested only in carrying specifications and not the whole verification, as for instance presented by Necula [3] in its Proof-Carrying Code technique. His solution is useful whenever the producer wants to state properties about its code. In our case instead, the security policy is defined by the code consumer.

JML - Java Modeling Language: The Java Modeling Language (JML) [2] is a behavioral interface specification language that can be used to specify the behavior of Java modules. It combines the design by contract approach of Eiffel [4] and the model-based specification approach of the Larch family of interface specification languages [5], with some elements of the refinement calculus.

Its use is strictly related to Java: the specifications are inserted directly into the source code as comments starting with the special sequence //@ or between /*@ and @*/. Such comments can be related to a class, a method, a field, a variable or a loop. Through a full set of special keywords (e.g. *requires, ensures*) it allows the programmer to specify *pre* and *post* conditions, invariants and many other constraints. It also allows to create "class models" that can be used for specification purposes. Finally, JML comes with a number of tools developed both by the JML team [6] (e.g. *jml-checker, jml-compiler*) and by third-parties (e.g. *ESC/Java2* [7]).

We choose *JML* because of its widespread support and strong structure, though it comes with some limitations:

- JML does not officially support Java 5. This is due to fact that JML tools released by the JML team are based upon the *MultiJava Project* [8] framework, a software for extending the Java programming language whose most recent release addressed Java 1.4. Third-party tools, like *ESC/Java2* [7] brought Java 5 support to JML, asserting the idea (that we also embrace) that this language is still able to grow. As we will see in Section 3.1, our approach is in fact based on language features introduced in Java 5, and this has led us to extend the JML grammar and the JML parser to take into account Java 5 extensions.

– JML is particularly addressed at static, source-level analysis: JML annotations are inserted into Java source as comments, thus making them unable to survive the compilation. In our approach, JML specifications are needed at run-time, even when (for efficiency or commercial reasons) the source code cannot be distributed. We translate JML comments into Java Annotations and let the Java compiler produce a Java class file including the specification as metadata.

JML into annotation: Annotations are a recent feature introduced in languages such as Java, C#, and other languages of the .NET family, which allow programmers to attach arbitrary, structured and type-safe meta-data to their code. These meta-data, in Java 5, are characterized by an identifier (akin to a class or interface name) and by a signature (or schema), akin to the fields of a class, where each field has an identifier and a value. Custom Annotation types are declared with a syntax similar to that of a class, through the @interface keyword. Only field of basic types, String, Class, Enum, Annotation, or arrays of the same are allowed, and default values for them can be defined in the declaration.

In a recent work [9], Taylor proposes three different approaches to encapsulate JML annotations into Annotations. We decide to follow and extend the *single annotation* approach, that appeared to us to be the more versatile w.r.t. future JML changes. Taylor's approach consisted in using an Annotation with a single value; such value would be the string as used in the JML annotated source. The example given in Figure 1 would be rewritten as in Figure 2.

The downside of Taylor's approach for our purposes is that the JML string still needs to be parsed at run-time, which imposes a huge performance penalty.

```
//@ requires vals.length > 0;
//@ ensures (\forall i; 0 <= i && i <= vals.length; \result <= vals[i] );
public int getMin( int [] vals ) { ... }
```

Fig. 1. An example of JML annotated method

```
@JML("requires vals.length > 0;")
@JML("ensures (\forall i; 0 <= i && i <= vals.length; \result <= vals[i] );")
public int getMin( int [] vals ) { ... }
```

Fig. 2. JML in annotations: Taylor approach

We avoid that by computing the JML AST at compilation-time, and using a source-to-source compiler to generate an equivalent Java source code, where the AST is serialized in binary form into annotations. This source is then passed to the Java Compiler to produce the final object class file (example in Figure 3).

```
@JML( {0x6D, 0x74, 0x63, 0x77, 0x6E, 0x73, 0x79, 0x6F, 0x75, ...} )
public int getMin( int [] vals ) { ... }
```

Fig. 3. JML in annotations: our approach

3.2 Code-Consumer Side

The approach we proposed in the Section 3.1 is related to the code-producer side of the framework. We showed how a producer can distribute a plug-in in an executable form which includes its own specification. Thanks to JML and Java Annotations we made such specifications available in a well structured manner to the run-time system. On the code-consumer side, the host application must now retrieve the specifications of needed plug-ins, verify them if needed, and finally match them against the local requirements.

Specification validation: Once the JVM loads the class files (viz. jar) for the plug-in, the framework is able to recover the module specification through the Java *Reflection* API. Furthermore, the framework exposes methods to inspect the JML AST. At this level, the specifications must be validated before use. The wrapper can be informed by the user about the protection level to use: (i) the user trusts the plug-in specifications, or (ii) the user does not.

In the former case the framework can skip the verification task and proceed to requirements gathering and matching. In the latter, we can perform static or dynamic analysis to ensure that the specification is trustable. Standard static analysis techniques can be applied, and in certain cases can even fully prove specification reliability. Alternatively, we can inject snippets of executable code in the bytecode stream for a class, to verify other properties at run-time. Generally, we can opt for a full dynamic analysis, or a combination of both.

The paradigmatic application of run-time checks is to generate the bytecode corresponding to the constraint expressed in a JML annotation, and to inject it directly into the bytecode of the object class (at the appropriate insertion point). This will make the plug-in run the check every time a method or part of it is executed. For example, what seen in Figure 1 would become, at run-time, the bytecode equivalent of the code in Figure 4.

There are few observations we have to do.

- If the normal execution of the module must be aborted for a failure of a JML-derived assertion, we throw a `RuntimeException()`. In fact, since `Runtime-Exception` is an unchecked exception, we can inject the bytecode without need to modify the method signature by adding a `throws` clause. Note that the modification of a method signature could lead to a failed verification by the bytecode verifier at link-time.

```
public int getMin( int [] vals ) {

  if( vals.length < 0 )
    throw new RuntimeException();

  int ret = vals[0];
  for( int i = 1; i < vals.length; i++ )
    if( vals[i] < ret )
      ret = vals[i];

  for( int i = 0; i < vals.length; i++ )
    if( vals[i] < ret )
      throw new RuntimeException();

  return ret;
}
```

Fig. 4. JML AST expansion; source code equivalent to injected bytecode from JML specifications is shown boxed

- JML helps with that due to the fact that it uses a subset of Java (plus a few keywords) for its expressions. Once the keywords are correctly interpreted by our framework to generate the right sequence of Java instructions, we obtain a fully Java compliant source code without further effort. To obtain the bytecode sequence ready for injection we can exploit directly the Java Compiler. Note that such compilation must be performed at run-time, to avoid meta-data corruption that could be still possible if performed at compile-time.
- Bytecode manipulation in Java is not provided by native API. Libraries like *ASM* [10] and *BCEL* [11] provide such functionalities. Our framework uses the *JDasm* library for bytecode manipulation [12], which was developed as part of the present research. Its main advantage is the transparency it offers in modifying and executing and already loaded class. In fact, as reported in [13], once a class is loaded into the *JVM*, its name is associated to a class definition that cannot be changed until the *JVM* instance ends. The techniques *JDasm* uses to overcome this limitation consists in sub-classing the given class C to create a brand new class C' which extends C. The modified method are reimplemented and through Java polymorphism C' can be used as C in the original context.

Requirement match: Once P's specification is validated, one last step is needed before granting execution to the plug-in. The requirements gathering and matching is the last stage of the loading process. In our idea the host application, the user and the system can specify their own requirements over P's behaviour. One possible way to do this is again using JML and Java Annotations: in fact, a traditional plug-in architecture sees a main application loading classes that implements a given interface. Such interfaces can be annotated with JML as well.

As for the specification validation, we apply static and dynamic analysis techniques to grant the P execution. First static analysis is used to reduce the number of checks needed at run-time. At this stage, the framework can find the plug-in unable to satisfy the requirements; in this case it will deny its execution. Otherwise, for all the required properties that are not determinable by simply looking at the specifications, the application needs to inject run-time checks. As described in Section 3.2 the requirement specified in JML can be compiled into a bytecode sequence ready to be injected. A survey of static analysis techniques that can be used to reduce the scope of run-time checks is found in [14].

4   Scenario

In the following we present a scenario of application for our framework. Consider a mobile application that implements a typical plug-in infrastructure to extend its functionalities. The application needs to access local files, including one (a standard Note on a hand-held device) where high confidential information is stored (i.e., credit card numbers), but for security reasons the user does not want to grant access to that file to any plug-in he is going to download from third-party (potentially unreliable) producers. On the other hand, these plug-ins can access other files containing less sensitive information.

This is a typical example where the producer of the plug-in cannot provide a proof that his code matches a user's requirements (that particular file is used only by our specific user), nor can the user trust the producer about a property the producer does not even know about. The plug-in may need to access the local file system at arbitrary locations, but the producer is unable to specify that a particular file would never be opened. In this case, the user, through JML, may declare a *model variable* f and an invariant I such that (`f instanceof File && f.equals(my_private_file)`) never holds during the plug-in life.

Helped by the module specification, the framework can inspect the plug-in executable to verify if in some case I is violated; if it is found that the value of f cannot be determined statically, it will inject bytecode for checking I at run-time.

5   Conclusions

In this paper we have presented an idea to enforce consumer-specified security properties for modular software. We have shown a way for the code-producer to attach behavioral specifications as meta-data directly into an object file using a combination of JML and Java Annotations. Once the specifications are available at run-time, we used bytecode injection techniques to verify both the specifications and the consumer security requirements.

We meant this to be innovative compared to the traditional techniques that wants a third-party authority or the code producer himself to be the correctness and security guarantor (i.e. like in Proof Carrying Code [3]). We have shown a simple scenario where the security requirements can not be foreseen by any authority or producer, since they are specific to a given final user. The idea should suggest that there is a plenty of possible scenarios where the standard proofs techniques (where properties are chosen by the producer of the plug-in) are unable to completely satisfy requirements that can vary from user to user, from application to application and from system to system.

The tool suite for the framework we have described is currently under development, but some of the producer-side facilities have already been completed. In particular we completed the source-to-source compiler to parse JML Java-5-compliant annotations and to translate them into standard Java Annotations. Also the bytecode manipulation library for code inspection and injection has been completed and already successfully used in several related applications. Future work includes applying static analysis to reduce the number of run-time checks injected, and extending the scope of the properties that can be expressed and handled by the framework.

References

1. Thompson, S.: A survey on model checking Java programs. Technical Report CSRG-407. Department of Computer Science, University of Toronto (2000)
2. Leavens, G., Cheon, Y.: Design by contract with JML (2003)
3. Necula, G.C.: Proof-carrying code. In: Proc. 24th ACM Symp. Principles of Programming Languages, pp. 106–119. ACM Press, New York (1997)
4. Thomas, P., Weedon, R.: Object-Oriented Programming in Eiffel, 2nd edn. Addison-Wesley, Reading (1997)
5. Guttag, J.V., Horning, J.H.: Larch: languages and tools for formal specification. Springer, Heidelberg (1993)
6. Leavens, G.T.: The Java modeling language (JML) home page, http://www.cs.ucf.edu/~leavens/JML/
7. Chalin, P., Kiniry, J.R., Leavens, G.T., Poll, E.: Beyond assertions: Advanced specification and verification with JML and ESC/Java2. In: de Boer, F.S., Bonsangue, M.M., Graf, S., de Roever, W.-P. (eds.) FMCO 2005. LNCS, vol. 4111, pp. 342–363. Springer, Heidelberg (2006)
8. Clifton, C., Millstein, T., Leavens, G.T., Chambers, C.: MultiJava: Design rationale, compiler implementation, and applications. ACM Transactions on Programming Languages and Systems 28(3) (2006)
9. Taylor, K.B., Rieken, J., Leavens, G.T.: Adapting the Java Modeling Language for Java 5 annotations. Technical Report 08-06, Department of Computer Science, Iowa State University (2008)
10. Bruneton, É., Lenglet, R., Coupaye, T.: ASM: a code manipulation tool to implement adaptable systems. In: Proceedings of the ASF (ACM SIGOPS France) Journées Composants 2002: Systèmes à composants adaptables et extensibles (Adaptable and extensible component systems) (2002)

11. Apache Jakarta Project: (BCEL - the bytecode engineering library),
    `http://jakarta.apache.org/bcel/`
12. Galilei, G.: JDasm, `http://jdasm.sourceforge.net`
13. Lindholm, T., Yellin, F.: The Java Virtual Machine Specification, 2nd edn. Prentice
    Hall PTR, Englewood Cliffs (2002)
14. Burdy, L., Cheon, Y., Cok, D., Ernst, M., Kiniry, J., Leavens, G.T., Rustan, K.,
    Leino, M., Poll, E.: An overview of JML tools and applications. International Jour-
    nal on Software Tools for Technology Transfer 7(3), 212–232 (2005)

Idea: Using System Level Testing for Revealing SQL Injection-Related Error Message Information Leaks

Ben Smith, Laurie Williams, and Andrew Austin

North Carolina State University, Computer Science Department
890 Oval Drive, Raleigh, NC, USA
{ben_smith,laurie_williams,andrew_austin}@ncsu.edu

Abstract. Completely handling SQL injection consists of two activities: properly protecting the system from malicious input, and preventing any resultant error messages caused by SQL injection from revealing sensitive information. The goal of this research is to assess the relative effectiveness of unit and system level testing of web applications to reveal both error message information leak and SQL injection vulnerabilities. To produce 100% test coverage of 176 SQL statements in four open source web applications, we augmented the original automated unit test cases with our own system level tests that use both normal input and 132 forms of malicious input. Although we discovered no SQL injection vulnerabilities, we exposed 17 error message information leak vulnerabilities associated with SQL statements using system level testing. Our results suggest that security testers who use an iterative, test-driven development process should compose system level rather than unit level tests.

Keywords: SQL, Exception, Tomcat, Java, web application, system level, unit testing, database, SQL injection attacks, coverage, error message.

1 Introduction

In this paper, we examine two input validation vulnerabilities that are in the CWE/SANS Top 25 Most Dangerous Programming Errors[1] due to their prevalence and potential damage: SQL injection vulnerabilities and error message information leak vulnerabilities. *SQL injection vulnerabilities* occur when a lack of input validation could allow a user to force unintended system behavior by altering the logical structure of a SQL statement using SQL reserved words and special characters [1, 2]. The CWE categorizes SQL injection vulnerabilities as a subset of *input validation vulnerabilities*, which occur when a system does not assert that input falls within an acceptable range, allowing the system to be exploited to perform unintended functionality [3]. *Error message information leak vulnerabilities* are caused when an application does not correctly handle an exceptional condition and, as a result, sensitive information is revealed to the attacker [4, 5]. We contend that in web

---

[1] The CWE/SANS Top 25 can be found at http://cwe.mitre.org/top25/.

F. Massacci, D. Wallach, and N. Zannone (Eds.): ESSoS 2010, LNCS 5965, pp. 192–200, 2010.
© Springer-Verlag Berlin Heidelberg 2010

applications, where security is paramount, input validation is comprised of *both* ensuring that input falls within an acceptable range (e.g. "integer") and that the application fails gracefully when input is *not* within said range.

To expose and mitigate SQL injection vulnerabilities at the white box level, a development team can execute unit tests that assert that malicious input is rejected by the components that communicate with the database [6]. In some development methodologies, components are constructed in horizontal slices that emanate from the ground up—the components that perform logic and interact with the database are composed and tested long before the user interface.   However, in an *iterative development methodology*, teams build software on a feature-by-feature basis in vertical slices that extend from the database to the user interface [22]. Additionally, *test-driven development* implies the incremental creation of tests throughout the development process [7].

The goal of this research is to assess the relative effectiveness of system and unit level testing of web applications to reveal both SQL injection vulnerabilities and error message information leakage vulnerabilities when used with an iterative test automation practice by a feature development team. We conducted a case study on four Java-based open source web applications: iTrust[2], Hispacta[3], LogicServices[4], and TuduLists[5]. In our case study, we executed and compared JUnit[6] unit tests and HtmlUnit[7] system level tests. The purpose of this study is to determine whether system level testing[8] could be used in an iterative or test-driven development scenario to expose both parts of input validation earlier in the lifecycle—an important component of building security in from the beginning [8].

The rest of this paper is organized as follows.   Section 2 presents the required background for understanding our study procedure.   After that, Section 3 describes the case study, including the subject applications and experimental setup.   Next, Section 4 presents the results of our case study. Section 5 presents limitations of the study. Finally, Section 6 describes the conclusions we reached from our study.

2 Background

In this section, we demonstrate an example of a SQL injection vulnerability and discuss error information leakage vulnerabilities.

SQL Injection Vulnerabilities. Consider a Java method used for the deletion of a patient's information in a medical record system. We present the relevant source

---

[2] http://sourceforge.net/projects/itrust

[3] http://sourceforge.net/projects/hispacta

[4] http://sourceforge.net/projects/logicservice

[5] http://sourceforge.net/projects/tudu

[6] http://www.junit.org

[7] http://htmlunit.sourceforge.net/

[8] The approach we propose in this paper tests the web application in the context of its server; a system level technique. However, our approach also targets specific areas ("hotspots") of the web application; a unit level technique. Thus, the way we use HtmlUnit in our case study is a hybrid of system level and unit level approaches, which is technically considered grey box testing [8, 9].

code for this operation in Figure 1 (assume that patients are deleted by their names). The vulnerability in this example was introduced in the line defining the SQL statement. The example we have presented in Figure 1 performs no input validation and, as a result, the example contains a SQL injection vulnerability relative to the use of the name parameter. An attacker could cause change to the interpretation of the SQL query by entering the SQL command fragment "` OR TRUE --" in the input field instead of any valid user name in the web form.

The single quotation mark (`) indicates to the SQL parser that the character sequence for the username column is closed, the fragment OR TRUE is interpreted as always true, and the fragment of the query after the hyphens (--) is a comment. The altered WHERE clause of the SQL statement will be interpreted as always true and thus every patient is deleted from the table. Because no input validation was performed, the attacker can exploit the system by inserting the malicious input in the name field, and cause truncation of the Patients table. Thus, the bolded statement in Figure 1 is an example of a SQL injection hotspot (or just "hotspot" in this paper)—any source code location that may contain a SQL injection vulnerability [1, 2].

```
...
java.sql.Connection mySQLConnector = DriverManager.getConnection();
java.sql.Statement s = mySQLConnector.createStatement("DELETE FROM
Patients WHERE Name = `" + name + "`;");
int result = s.executeUpdate();
        return 1 == result;
...
```

Fig. 1. Patient Deletion Code in Java; hotspot is bolded

Error message information leak vulnerabilities. These vulnerabilities occur when an application does not correctly handle exceptional conditions and subsequently leaks sensitive information to a user [4, 5]. This information can be obviously dangerous in the case of error messages that contain system or application passwords, or it may seem more benign, containing only version numbers or stack traces. Unfortunately, even these seemingly benign error information leaks can provide valuable information to an attacker and could expose additional attack vectors. Since a tester cannot tell what information an attacker needs to conduct future attacks, a good policy is to treat all error information leakage vulnerabilities as if they contain obviously dangerous information such as passwords.

3   Case Study

In this section, we present information about our case study. Each part of the case study was conducted using Eclipse v3.3 Europa executed using Java v1.6 running on an IBM Lenovo T61p running Windows Vista Ultimate with a 2.40Ghz Intel Core Duo processor and 2GB of RAM. We used MySQL[9] v5.0.45-community-nt for our research database management system.

---

[9] http://www.mysql.com

Table 1. Information about the Test Subjects (n=4)

Project	iTrust	Hispacta	LogicServices	TuduLists
Version	4.0	0.0.3	1.8	2.2
Lines of Code[d]	7707	1991	5011	6178
Production Classes[*]	143	42	155	132
Database Classes	20	4	1	5
Hibernate[10]	No	Yes	Yes	Yes

[d] Source Lines of Code calculated by NLOC: http://sourceforge.net/projects/nloc/
[*] A production class is any class that is required to be on the class path for the web application to function correctly (excluding test classes and utility classes)

To obtain our case study applications, we collected information about 12 enterprise Java web applications, which we found by searching SourceForge[11] with the query "Java web application," and sampling the first 12 projects that contained the Eclipse webtools[12] project file structure. We then rejected eight subjects from our study because they did not meet one or many of the following criteria:

- Could be compiled, built, and deployed.
- Contained automated unit tests, written in JUnit, which were distributed with the source code.
- Relied upon a relational DBMS to store its data.

We were left with the four subjects presented in Table 1. In an attempt to reveal both SQL injection and error message information leak vulnerabilities in our test subjects, as stated in Section 1, we created the following systematic, system level, security testing procedure and executed it on our subjects. By design, this procedure produces an automated system level test suite that executes all reachable hotspots with normal and malicious input. We note here that by *intrinsic*, we mean that we did not augment or modify the existing test set in any way; values for these measures were achieved by the unit tests that were distributed with each system.

1. **Identify and Instrument Hotspots.** We manually inspected the source code to discover any point where the system interacts with the database. We note here that hotspots can take many forms; we explain this issue more below. We have written the Java program SQLMarker, introduced in our earlier work [9]. SQLMarker keeps a record of the execution state at runtime for each uniquely identified hotspot[13]. SQLMarker has a method, SQLMarker.mark(), which passes the line number and file name to a research database that stores whether the hotspot has been executed.

---

[10] http://www.hibernate.org/
[11] http://sourceforge.net/
[12] http://www.eclipse.org/webtools/
[13] For larger applications, one could use a static analyzer to determine hotspots' locations.

2. **Record Hotspots.** A second class we wrote, called `Instrumenter`, provides each manually marked hotspot with a unique identifier comprised of the filename and line number, and outputs the number of hotspots found. Once we manually marked each, we executed `Instrumenter` to store a record of each of these hotspots.

3. **Execute Original Unit Tests.** After instrumenting each subject to mark its executed SQL hotspots, we executed the intrinsic unit tests and recorded the resultant number of executed statements.

4. **Create Test Cases.** We used the stored file name and line number of the hotspot from Step 1 to construct an automated system level test with HtmlUnit[14] that executed the SQL statement located at the stored file name and line number. We constructed an initial automated test for each hotspot by using a call hierarchy and manual testing to make web requests until the hotspot was marked as being executed and then modeled our automated test after the use case we discovered[15].

5. **Apply Malicious Input.** We modified the test defined in Step 4 to emulate a malicious user by using 132 forms of malicious input in an attack list from NeuroFuzz [10] in place of normal input. This part of the procedure is similar to "fuzzing". The difference here is that fuzzing is a semi-random, black box activity; our approach is targeted to specifically attack the areas where user input might reach a hotspot.

6. **Record Result.** We then marked each test that caused incorrect SQL operations or an application error in Step 5 as a successful attack and its corresponding SQL statement as a vulnerability.

Identifying hotspots may seem trivial, but in fact can be difficult because hotspots may not always take the same form. One way of discovering vulnerabilities is automated static analysis tools, which can be designed to check for a particular hotspot or vulnerability type [11]. We executed static analysis tools on our subjects and the tools reported no input validation vulnerabilities in any project we examined.

4 Results

This section presents the results of our case study. We first observed, as shown in Table 2, that there were no intrinsic JUnit test cases that used malicious input. We conducted all of our 272 system level tests by using HtmlUnit to inject our attack list into a request parameter, or in the case of TuduLists, to conduct an AJAX request where the malicious input was injected into an asynchronous JavaScript call. Using our technique, we found no instances of SQL injection vulnerabilities at the system

---

[14] HtmlUnit is a "GUI-Less browser for Java programs". It models HTML documents and provides an API that allows you to invoke pages, fill out forms, click links, etc, just like you do in your "normal" browser. http://htmlunit.sourceforge.net.

[15] However, some hotspots were not used by the JSPs in the application, perhaps because these hotspots were used for database administration only, or the development team had not finished implementing the use case that required the query. If we could not reach the SQL statement through the web interface, we augmented the white box test plan to include a malicious test that directly calls the database class.

level. No application allowed us to issue commands to the database management system because prepared statements and Hibernate both perform strong type checking on the variables used in their hotspots. Hibernate allows developers to create persistent classes in the object-oriented paradigm that represent individual database records [12]. However, we found 17 error message information leak vulnerabilities among the four applications in our case study, summarized in Table 2.

Table 2. Results for the Test Subjects

Project	iTrust	Hispacta	LogicServ ices	TuduLists
Hotspots	92	23	48	13
Covered by Intrinsic Tests	89	20	47	3
Statement Coverage (EclEmma)	84%	49%	53%	40%
Test Cases with Malicious Input	0	0	0	0
New System Level Test Cases (Normal and Malicious)	149	29	80	14
Confirmed Vulnerabilities	2	2	9	4

We found that unit testing could have identified none of the 17 confirmed vulnerabilities; rather, these confirmed vulnerabilities are system level vulnerabilities that had to involve the application server. A missing exception handler for pages or Servlets within Apache Tomcat caused each of the vulnerabilities we discovered. The example presented in Table 3 and Table 4 helps illustrate why these vulnerabilities cannot be exposed at the unit level.

Table 3 presents the relevant code from one of the confirmed vulnerabilities that we found in iTrust, in the file editHCPs.jsp. In other pages within iTrust, there is a JSP directive declared at the top of the page's code (along with various navigational toolbars and headers) that declares an exception handler:

```
<%@page errorPage="/auth/exceptionHandler.jsp"%>
```

This directive does not appear in editHCPs.jsp (see Table 3). At the moment an exception is thrown, Apache Tomcat forwards the user to the page declared in this directive, if this directive is declared. Otherwise, Apache Tomcat outputs a revealing stack trace to the user's browser window, also known as an error message information leakage.

Since the omission of an exception handler is something that happens in the JSP code and not the Java code, some form of interaction is required with the application server (Apache Tomcat) in order to expose the vulnerabilities. One may view each JSP as a unit, but still the exception handler is a JSP page directive that involves a separate page; the unit therefore cannot be tested in isolation. The confirmed

vulnerabilities, then, are caused by a system level error: the absence of an exception handler in the JSP or Servlet code of the application. Consider a JUnit test case that is written to execute undeclareHCP (see Table 4). This JUnit test case would pass, but would not expose the vulnerability even if it uses the some malicious input, such as '
UNION SELECT. However, an HtmlUnit test case that targets editHCPs.jsp (see Table 3), produced by our system level testing technique, would expose the vulnerability using the same attack. That is, the vulnerability is not that an exception is thrown, but rather that the exception is not correctly handled by the JSP.

Table 3. JSP for Example Vulnerability (editHCPs.jsp)

```
DeclareHCPAction action = // Action class for declaring the HCP.
String confirm = ""; // used to store the result from the DAO.
String removeHCP = request.getParameter("removeID");
if(removeHCP!=null && !removeHCP.equals("")){
        confirm = action.undeclareHCP(removeHCP);
}
List<PersonnelBean> hcps = action.getDeclaredHCPS();
```

Table 4. Java Method undeclareHCP

```
//given: patientDAO, a DAO pertaining to the patients table
//given: iTrustException, a custom-build Exception class for
//        handling alternate flow errors
public String  undeclareHCP(String input) throws iTrustException {
try {
long hcpID = Long.valueOf(input);
boolean confirm = patientDAO.undeclareHCP(loggedInMID, hcpID);
if (confirm) {
        return "HCP successfully undeclared";
} else
        return "HCP not undeclared";
} catch (NumberFormatException e) {
        throw new iTrustException("HCP's MID not a number");
} catch (DBException e) {
        throw new iTrustException(e.getMessage());
}
}
```

5   Limitations

Future case studies should examine much larger web applications than the ones in this study. In addition, the selective criteria as described in Section 3 could have biased the data. For example, perhaps the fact that all of our test subjects were Tomcat Servlet applications caused or prevented some security vulnerabilities that would not have been observable in another architectural setup. In addition, if stored procedures had been used in any of our test subjects, our results may have been different. The development teams for each project may have been using other testing techniques to improve the security posture of our subjects, or security may not have been high on their list of requirements.

The container for the applications (in this case, Apache Tomcat) could also be emulated using a Mock Object pattern [13], and each individual servlet or JSP could be tested in isolation from one another. However, the quality of the testing results is entirely dependent on the quality of the mock object's ability to emulate the server [13]; additionally, mock objects may not be any less expensive than system testing. Prepared statements, which separate the user's input from the structure of the query at the application level [14], protected the applications in this study. However, prepared statements are only useful if developers are aware of them and choose to use them. Our own system level procedure may not have exposed all vulnerabilities latent in the four subjects. Our procedure was targeted towards SQL injection vulnerabilities, which did not exist in these sampled applications at the locations of the hotspots we identified, but other vulnerabilities of varying types may exist in our subjects.

6 Conclusion

In our investigation of the relative effectiveness of unit and system level testing techniques, we have discovered that developers sometimes miss the fact that input validation is comprised of *both* ensuring that input falls within an acceptable range (e.g. "integer") and that the application fails gracefully when input is *not* within said range. We found that all four of our study subjects use Hibernate and/or properly constructed prepared statements, which were completely effective for asserting that input falls within a safe (non-attack) range. Using a systematic system level security testing procedure to generate an HtmlUnit test suite, we found 17 error message information leakage vulnerabilities in the four web applications of our study. We found it impossible to replicate these same 17 vulnerabilities by augmenting the intrinsic unit test suites with additional malicious tests because vulnerabilities cannot be exposed at the system level though unit testing.

Our results show that ensuring that error messages resulting from SQL injection attacks do not reveal sensitive information is an inherently system level activity because the web server will dictate how and when error messages are displayed. *Thus, an iterative, a feature-based development team conducting a test-driven automation practice can use a system level test procedure like the one described in this paper to expose both SQL injection vulnerabilities and error message information leak vulnerabilities.* From a security perspective, unit testing would not be effective toward this aim, because it cannot take into account the production environment in which the system exists.

Acknowledgments. We would like to thank the North Carolina State University Realsearch group for their helpful comments on the paper. In addition, we would like to thank Yonghee Shin for the foundational work she performed by providing formal definitions for our SQL hotspot metrics and for her input on the content of this paper. This work is supported by the National Science Foundation under CAREER Grant No. 0346903. Any opinions expressed in this material are those of the author(s) and do not necessarily reflect the views of the National Science Foundation.

References

1. Halfond, W.G.J., Orso, A.: AMNESIA: analysis and monitoring for neutralizing SQL-injection attacks. In: 20th IEEE/ACM International Conference on Automated Software Engineering, Long Beach, CA, USA, pp. 174–183 (2005)
2. Kosuga, Y., Kono, K., Hanaoka, M., Hishiyama, M., Takahama, Y.: Sania: syntactic and semantic analysis for automated testing against SQL injection. In: 23rd Annual Computer Security Applications Conference, Miami Beach, FL, pp. 107–117 (2007)
3. Pietraszek, T., Berghe, C.V.: Defending against injection attacks through context-sensitive string evaluation. In: Valdes, A., Zamboni, D. (eds.) RAID 2005. LNCS, vol. 3858, pp. 124–145. Springer, Heidelberg (2006)
4. Aslam, T., Krsul, I., Spafford, E.: Use of a taxonomy of security faults. In: 19th National Information Systems Security Conference, Baltimore, MD, pp. 551–560 (1996)
5. Tsipenyuk, K., Chess, B., McGraw, G.: Seven pernicious kingdoms: a taxonomy of software security errors. IEEE Security & Privacy 3, 81–84 (2005)
6. IEEE: IEEE Standard 610.12-1990, IEEE Standard Glossary of Software Engineering Terminology (1990)
7. Beck, K.: Test-driven development: By example. Addison-Wesley, Boston (2003)
8. McGraw, G.: Software security: Building security in. Addison-Wesley, Upper Saddle River (2006)
9. Smith, B., Shin, Y., Williams, L.: Proposing SQL statement coverage metrics. In: The 4th International Workshop on Software Engineering for Secure Systems at the 30th International Conference on Software Engineering, Leipzig, Germany, pp. 49–56 (2008)
10. Jiang, Y., Cukic, B., Menzies, T.: Fault Prediction using Early Lifecycle Data. In: The 18th IEEE International Symposium on Software Reliability, 2007. ISSRE 2007, pp. 237–246 (2007)
11. Livshits, V.B., Lam, M.S.: Finding security vulnerabilities in Java applications with static analysis. In: USENIX Security Symposium, Baltimore, MD, pp. 18–18 (2005)
12. Bauer, C., King, G.: Hibernate in Action. Manning Publications (2004)
13. Brown, M., Tapolcsanyi, E.: Mock object patterns. In: The 10th Conference on Pattern Languages of Programs, Monticello, USA (2003)
14. Thomas, S., Williams, L.: Using automated fix generation to secure SQL statements. In: Proceedings of the Third International Workshop on Software Engineering for Secure Systems, Minneapolis, MN (2007)

Automatic Generation of Smart, Security-Aware GUI Models

David Basin[1], Manuel Clavel[2,3], Marina Egea[1], and Michael Schläpfer[1]

[1] ETH Zürich, Switzerland
{basin,marinae,michschl}@inf.ethz.ch
[2] IMDEA Software Institute, Madrid, Spain
manuel.clavel@imdea.org
[3] Universidad Complutense de Madrid, Spain

Abstract. In many software applications, users access application data using graphical user interfaces (GUIs). There is an important, but little explored, link between visualization and security: when the application data is protected by an access control policy, the GUI should be aware of this and respect the policy. For example, the GUI should not display options to users for actions that they are not authorized to execute on application data. Taking this idea one step further, the application GUI should not just be security-aware, it should also be smart. For example, the GUI should not display options to users for opening other widgets when these widgets will only display options for actions that the users are not authorized to execute on application data. We establish this link between visualization and security using a model-driven development approach. Namely, we define and implement a many-models-to-model transformation that, given a security-design model and a GUI model, makes the GUI model both security-aware and smart.

1 Introduction

In many programs, users access application data using GUI widgets: data is created, deleted, read, and updated using text boxes, check boxes, buttons, and the like. There is an important, but little explored, link between visualization and security: When the application data is protected by an access control policy, the application GUI should be *aware of* and *respect* this policy. For example, the GUI should not display options to users for actions that they are not authorized to execute on application data. This prevents user frustration, for example, from filling out a long electronic form only to have the server reject it because the user lacks a permission to execute some associated action on the application data. Taking this idea one step further, the GUI should not, for example, display options to users to open other widgets when these widgets only display options for actions that the users are not authorized to execute on application data. That is, the application GUI should not just be security-aware but also *smart*.

F. Massacci, D. Wallach, and N. Zannone (Eds.): ESSoS 2010, LNCS 5965, pp. 201–217, 2010.

Visualization and security

To see how this link between GUIs and security policies might look, consider the following example: an application for managing employee information. This information will include, among other data, employees' names, phone numbers, and salaries. Suppose now that the employee information is protected by an access control policy that includes, among other clauses, the following:

- All users can read employees' names.
- Only administrators and supervisors can read and update employees' phone numbers.
- Only supervisors can read employees' salaries.

Suppose now that, as shown in Figure 1, our application GUI includes the following windows:

Window #1. This is our main window. Here, users can edit employee data by clicking on the Edit Phone Number-button.

Window #2. Users can select an employee from a list of names, shown in a combo box, and view the information associated to the selected employee by clicking on the View-button.

Window #3. Users can read in the Name, Phone, and Salary-labels, respectively, the name, phone number, and salary of the selected employee. Moreover, users can edit the phone number by clicking on the Edit-button.

Window #4. Users can update the phone number of the selected employee by typing the new number into the Phone-entry and clicking on the OK-button.

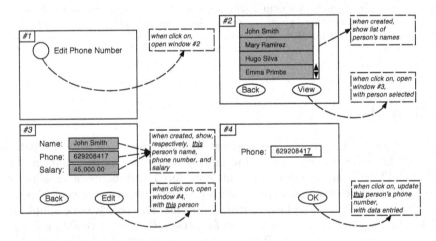

Fig. 1. A simple GUI for editing employees' phone numbers

What behaviour should our GUI have if it is to be considered "security-aware"? Suppose that a user with the role administrator wants to edit an employee's phone number using our GUI. Since administrators are not authorized to read employees' salaries, when opening Window #3, our GUI should prevent an administrator from reading this information in the Salary-label. Furthermore, how should our GUI behave if it is also to be considered "smart"? Suppose now that a user with no special privileges wants to edit an employee's phone number using our GUI. Since ordinary users are not authorized to read or update employees' phone numbers, our GUI should prevent the user from opening Window #4 by clicking on the Edit-button, since the user should not be able to do anything within this window (i.e., neither read the phone number of the selected employee in the Phone-label nor click on the OK-button to update this information).

The problem we address here is how to establish this link between visualization and security. The default, "ad-hoc" solution, namely, directly hardcoding the security policy within the GUI, is clearly inadequate. First, the GUI designer is often not aware of the application data security policy. Second, even if the designer is aware of it, hardcoding the application data security policy within the GUI code is cumbersome and error-prone, if done manually. Finally, any changes in the security policy will require manual changes to the GUI code where this policy is hardcoded, which again is a cumbersome and error-prone task.

Our approach: model-transformation

We propose in this paper a model-driven approach that links visualization and security. The key idea is that this link is ultimately defined in terms of *data actions*, since data actions are both controlled by the security policy and triggered by the events supported by the graphical user interface. The key component of our proposal is a many-models-to-model transformation which, given a security-design model (specifying the access control policy on the application data) and a GUI model (specifying the actions triggered by the events supported by the application's graphical interface), automatically generates a GUI model that is both security-aware and smart. Thus, under our proposal, illustrated in Figure 2, the process of modeling a smart, security-aware GUI has the following parts.

1. Software engineers specify the application-data model $\mathcal{C}$.
2. Security engineers specify the security-design model $\mathcal{S}_\mathcal{C}$.
3. GUI designers specify the application GUI model $\mathcal{G}_\mathcal{C}$.
4. A many-models-to-model transformation automatically generates a smart, security-aware GUI model $\mathcal{M}_{(\mathcal{G}_\mathcal{C}, \mathcal{S}_\mathcal{C})}$ from the security model $\mathcal{S}_\mathcal{C}$ and the GUI model $\mathcal{G}_\mathcal{C}$.

To show the applicability of our proposal, we have implemented an Eclipse-based application that automatically generates smart, security-aware GUI models from security-design models and GUI models [7]. Moreover, it automatically generates smart, security-aware web applications from the generated smart, security-aware GUI models. Specifically, our Eclipse-application includes the following parts.

- A GMF editor for drawing application-data models.
- A plugin for generating user-friendly GMF editors for drawing security-design models and GUI models.

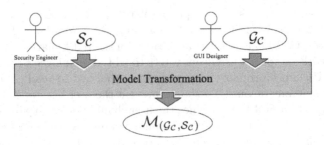

Fig. 2. Modeling a smart and security-aware GUI

- A plugin for generating smart, security-aware GUI models from security-design models and GUI models.
- A plugin for generating web applications from smart, security-aware GUI models.

Due to space limitations, we only report on our plugin for generating smart, security-aware GUI models. This plugin implements our many-models-to-model transformation as a QVT operational transformation.

Applicability and extensions

The applicability of our approach crucially depends on the expressiveness of the modeling languages used to specify the access control policies and the graphical user interfaces. In this paper, however, we focus on the two main ideas behind our approach.

1. The link between visualization and security is essentially given by the data actions, since they are both controlled by the security policy and triggered by the events supported by the graphical user interface.
2. This link can be systematically established using an appropriate many-models-to-model transformation to generate smart, security-aware GUI models from the models specifying the security policy and the models specifying the graphical user interfaces.

To explain these ideas, we present our approach using abstract notions of both security-design models and (smart, security-aware) graphical user interface models. For the sake of illustration, we will also use concrete modeling languages, which provide the source and target models of a many-models-to-model transformation that we will introduce to exemplify our approach; however, our approach is not restricted to or dependant on the use of these languages or our particular many-models-to-model transformation. In fact, our Eclipse-based application for generating smart, security-aware GUIs [7] currently supports a modeling language for specifying graphical user interfaces that is significantly more expressive than the one introduced in this paper. For example, it allows one to associate data to widgets, to pass information from one widget to another widget, to jump from one widget to another widget, and to call actions on data with parameters.

Notice that some of these features are, for example, needed to fully modeled the graphical user interface described in Figure 1.

Organization. In Sections 2 and 3 we introduce security-design models and GUI models. Afterwards, in Sections 4 and 5, we introduce smart, security-aware GUI models and we define a many-models-to-model transformation that automatically generates smart, security-aware GUI models from security-design models and GUI models. We conclude with a discussion of related and future work. Throughout the paper we will use the employee information system example, given above, as our running example.

2   Security-Design Models

Model-driven security (MDS) [2] is a specialization of model-driven development for developing secure systems. In this approach, designers specify system models along with their security requirements and use tools to automatically generate system architectures from the models, including complete, configured access control infrastructures. MDS is centered around the construction (and analysis) of *security-design models*, which are models that combine security requirements with system designs.

In this section, we define security-design models as they will be considered throughout this paper. Our focus is on access control security requirements. We first provide an abstract definition (Definition 2) of security-design models, independent of any modeling language that may be used for specifying them. Then, we introduce a specific language (SecureUML+ComponentUML) for modeling security-design models.

We begin by defining system design models (Definition 1) and by introducing a specific modeling language for them (ComponentUML). For the sake of simplicity, we will consider that system designs are component-based, and we will use the following (rather simple) notion of a component-based design model throughout this paper

Definition 1. *A* component-based design model $\mathcal{C}$ *is a 4-tuple*

$$\mathcal{C} = \langle E, At, As, Md \rangle$$

that specifies the entities E which play a role in the system as well as their properties, given by their attributes At, associations-ends As, and their methods Md.

Example 1. The component-based design model specifying the data model underlying our running example will consist of a single entity (*Person*), with three attributes (*name, phone number,* and *salary*).

The ComponentUML language. ComponentUML is a simple language for modeling component-based systems. Essentially, it provides a subset of UML class models: entities can be related by associations and may have attributes and methods. Its metamodel is shown in Figure 3 (inner rectangle).

Each valid instance of the ComponentUML metamodel specifies a component-based design model $\langle E, At, As, Md \rangle$ whose components are defined by the following OCL expressions:

E = Entity.allInstances().
At = Attribute.allInstances().
As = AssociationEnd.allInstances().
Md = Method.allInstances().

We are now ready to define security-design models.

Definition 2. *Let $\mathcal{C}$ be a component-based model $\mathcal{C} = \langle E, At, As, Md \rangle$. Then, a security-design model $\mathcal{S}_\mathcal{C}$ for $\mathcal{C}$ is a 5-tuple*

$$\mathcal{S}_\mathcal{C} = \langle Rs, DaAc, Rl, RsDaAc, DaAu \rangle,$$

with $Rs = (E \cup At \cup As \cup Md)$, $RsDaAc : Rs \longrightarrow \mathcal{P}(DaAc)$, and $DaAu : DaAc \longrightarrow \mathcal{P}(Rl)$. The model $\mathcal{S}_\mathcal{C}$ specifies a security policy for accessing the resources (namely, the entities and their properties) in the component-based system modeled by $\mathcal{C}$. More concretely, $\mathcal{S}_\mathcal{C}$ specifies

- *the actions $DaAc$ whose access policy is modeled;*
- *the specific actions $RsDaAc(rs)$ supported by a given resource $rs \in Rs$;*
- *the roles Rl that users may adopt when interacting with the system; and*
- *the roles $DaAu(daac) \subseteq Rl$ that are authorized to execute a given action $daac \in DaAc$.*

Example 2. The security-design model specifying the security policy of our running example will define, for example, that:

- the roles that users may adopt when using the system are *all* (for users with no special privileges), *administrator*, and *supervisor*;
- the actions supported by the resources include, for example, to *read* and *update* the *phone number* resource; and
- the role *supervisor*, for example, is authorized to execute a *read* action on (any) *salary* resource, but not the other two roles.

The SecureUML language. This is a modeling language based on RBAC [5] for formalizing access control policies on protected resources [2]. The policies that can be specified in SecureUML are of two kinds: those that depend on static information, namely the assignments of users and permissions to roles and those that depend on dynamic information. SecureUML leaves open what the protected resources are and which actions they offer to clients. These are specified in a so-called *dialect* and depend on the primitives for constructing models in the associated system-design modeling language. Each SecureUML dialect basically declares its own protected resources and the actions that they offer to clients.[1]

The SecureUML+ComponentUML language. This is a SecureUML dialect that connects SecureUML with ComponentUML, providing a convenient language for specifying security-design models. Its metamodel is shown in Figure 3.

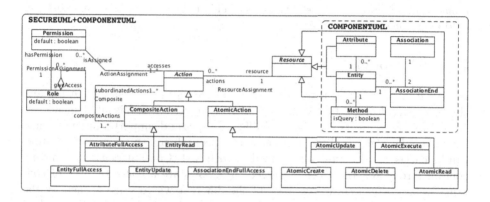

Fig. 3. SecureUML+ComponentUML metamodel

The protected resources are the entities, as well as their attributes, methods, and association-ends. The atomic actions that are offered to clients are create, delete, update, read, and execute the entity's properties or methods. The dialect also provides composite actions, which are used to group primitive actions into a hierarchy of higher-level ones. The composite actions that are offered to clients are read, update, and full access either on entities or entity's properties: e.g., full access on an attribute includes both read and update access on this attribute.

Each valid instance[2] of the SecureUML+ComponentUML metamodel specifies a security-design model $S_C = \langle Rs, DaAc, Rl, RsDaAc, DaAu \rangle$, whose components are defined by the following OCL expressions:

$Rs = $ Resource.allInstances().
$DaAc = $ AtomicAction.allInstances().
$Rl = $ Role.allInstances().
$RsDaAc(rs) = rs$.actions−>select(a|a.oclIsTypeOf(AtomicAction)).
$DaAu(daac) = daac$.allAssignedRoles().

---

where the operation allAssignedRoles() is defined as follows:[3]

context AtomicAction::allAssignedRoles():Set(Roles)
body: self.compactionPlus().isassigned.allRoles()−>asSet()

Example 3. Suppose that EmployeeSalaryAtomicRead denotes the action of reading (any) *salary* resource and that Supervisor denotes the role *supervisor*. Then, for any instance of the SecureUML+ComponentUML metamodel that correctly models the access control policy in our running example, $DaAu$(EmployeeSalaryAtomicRead) should return Set{Supervisor}. This is because only supervisors can read employees' salaries.

3   GUI Models

In GUI design, it is useful to distinguish between a GUI's visual elements and the behavioural properties associated with these elements. Visual elements are typically called widgets, which are of different types and support different events. Examples of widgets include buttons, entries, containers, and windows. Buttons can be clicked on. Entries can be filled in with text. Containers can graphically contain (group together) other widgets. Windows are a concrete class of containers. The behavioural properties of the different widgets are defined by the actions associated to the events that they support. We distinguish between data actions and widget actions. Data actions act on application data. Widget actions act upon widgets (including themselves).

In this section, we define GUI models as they will be considered throughout this paper. We first provide an abstract definition and afterwards we introduce a specific language for modeling GUIs.

Definition 3. *Let C be a component-based model $C = \langle E, At, As, Md \rangle$. Then, a GUI model $\mathcal{G}_C$ for C is a 9-tuple[4]*

$$\mathcal{G}_C = \langle Wd, Wd^c, In, Ev, DaAc, WdAc, WdEv, EvWdAc, EvDaAc \rangle,$$

with $Wd^c \subseteq Wd$, $In : Wd^c \longrightarrow \mathcal{P}(Wd)$, $WdEv : Wd \longrightarrow \mathcal{P}(Ev)$, $EvWdAc : Ev \longrightarrow \mathcal{P}(WdAc)$, and $EvDaAc : Ev \longrightarrow \mathcal{P}(DaAc)$. The model $\mathcal{G}_C$ specifies a graphical interface to interact with the component-based system modeled by C. More concretely, $\mathcal{G}_C$ specifies

---

[3] The auxiliary operations compactionPlus() and allRoles() return, respectively, the collection of composite actions to which an action is (directly or indirectly) subordinated and the collection of roles that are (directly or indirectly) assigned to a given permission. We again refer to [1] for the full definitions of these operations.

[4] For the sake of simplicity, we have left implicit the intended dependency of $\mathcal{G}_C$ with respect to C: namely, that the data actions $DaAc$ are indeed actions upon the entities (and their properties) that are modeled in C. To make this dependency explicit, similarly to what we did for the case of security-design models, we would extend our 9-tuple in Definition 3 with two additional components: namely Rs and $RsDaAc$, with $Rs = (E \cup At \cup As \cup Md)$ and $RsDaAc : Rs \longrightarrow \mathcal{P}(DaAc)$.

- *the widgets Wd and widget containers Wd^c that make up the GUI;*
- *the widgets In(wd) that are contained by a given widget container wd;*
- *the events Ev supported by the GUI;*
- *the data actions DaAc and the widget actions WdAc that can be triggered by the events;*
- *the events WdEv(wd) $\subseteq$ Ev supported by a given widget wd $\in$ Wd; and,*
- *the widget actions EvWdAc(ev) $\subseteq$ WdAc and data actions EvDaAc(ev) $\subseteq$ DaAc associated with a given event ev $\in$ Ev.*

Example 4. The GUI model (partially) specifying the graphical user interface in our running example will, for example, define that: the widgets are four windows, which contain six buttons, one combo-box, three labels and one entry; the buttons support, among others, *on click* events; and the *on click* event supported by the View-button triggers the widget action of opening the Window #3, while the *on click* event supported by the OK-button triggers the data action of *updating* a given *phone number* resource.

The GUI language. This is a simple language for modeling GUIs. The GUI metamodel is shown in Figure 4 (inner rectangle).[5]

Application GUIs consist of widgets that are displayed inside containers, which are themselves widgets. Each widget has a (possibly empty) set of events associated to it, and each event is in turn associated with a set of actions, which are the actions triggered by the event. Also, the events' actions are of two types: widget actions (which are actions on GUI widgets) and model actions (also denoted data actions), which are actions on the application data.

Each valid instance of the GUI metamodel specifies a GUI model $\mathcal{G}_\mathcal{C} = \langle Wd, Wd^c, In, Ev, DaAc, WdAc, WdEv, EvWdAc, EvDaAc \rangle$, whose components are defined by the following OCL expressions:

Wd = Widget.allInstances().
Wd^c = Container.allInstances().
$In(wd)$ = wd.contained.
Ev = Event.allInstances().
$DaAc$ = DataAction.allInstances().
$WdAc$ = WidgetAction.allInstances().
$WdEv(wd)$ = wd.widgetEvents.
$EvWdAc(ev)$ = ev.firedActions−>select(a|a.oclIsTypeOf(WidgetAction)).
$EvDaAc(ev)$ = ev.firedActions−>select(a|a.oclIsTypeOf(ModelAction))
                            .oclAsType(ModelAction).modelAction).

---

[5] For the sake of simplicity, our metamodel only defines a basic subsets of widgets (windows, entries, and buttons), of events (entering or leaving a widget, creating a widget, and clicking or double-clicking on a widget), and of widget actions (opening and closing a widget). These subsets are, however, sufficient for the purpose of this paper. The interested reader can find in [7] the definition of a more comprehensive GUI metamodel.

Example 5. Suppose that onCreateSalaryLabelWindow3 denotes the event that creates a Salary-label in Window #3. Then, for any instance of the GUI metamodel that correctly models the graphical user interface in our running example, *EvDaAc*(onCreateSalaryLabelWindow3) should return Set{EmployeeSalaryAtomic-Read}. This is because reading a *salary* resource is the data action triggered when the Salary-label is created.

4   Security-GUI Models

We are now ready to provide an abstract definition (Definition 4) of security-GUI models. These are models in which permissions are associated to widget events in order to specify who can execute them. We also introduce a specific language for modeling security-GUI models (SecureUML+GUI).

Definition 4. *Let C be a component-based model $C = \langle E, At, As, Md \rangle$. Let $\mathcal{G}$ be a GUI model for C*

$$\mathcal{G}_C = \langle Wd, Wd^c, In, Ev, DaAc, WdAc, WdEv, EvWdAc, EvDaAc \rangle.$$

Then, a security-GUI model $\mathcal{M}_{\mathcal{G}_C}$ for $\mathcal{G}_C$ is a triple $\mathcal{M}_{\mathcal{G}_C} = \langle \mathcal{G}, Rl, EvAu \rangle$, with $EvAu : Ev \longrightarrow \mathcal{P}(Rl)$, that specifies a security policy for accessing the resources (namely, the events) in the graphical interface modeled by $\mathcal{G}_C$. More concretely, it specifies all the roles Rl that users may adopt when interacting with the GUI, as well as the specific roles $EvAu(ev) \subseteq Rl$ that are authorized to execute a given event $ev \in Ev$.

Example 6. The security-GUI model specifying the expected (security) behaviour of the graphical user interface in our running example will, for example, define that the role *administrator* is authorized to execute the events creating the Name and Phone-labels in Window #3, but not the event creating the Salary-label.

The SecureUML+GUI language. SecureUML+GUI is another dialect of SecureUML. It combines SecureUML with our simple GUI modeling language, providing a convenient language for specifying security-GUI models.

The SecureUML+GUI metamodel, shown in Figure 4, provides the connection between SecureUML and GUI. It specifies the protected resources, namely, events, as well as the available actions on these protected resources, namely, their execution.

Each valid instance of the SecureUML+GUI metamodel specifies a security-GUI model $\mathcal{M}_{\mathcal{G}_C} = \langle \mathcal{G}_C, Rl, EvAu \rangle$, whose components are defined by the following OCL expressions:

$Rl =$ Role.allInstances().
$EvAu(ev) = ev$.allAssignedRoles().

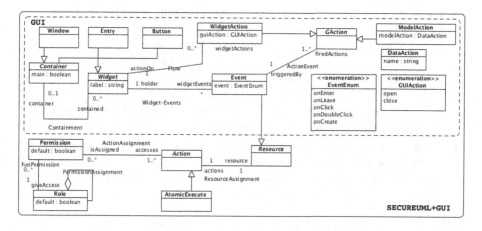

Fig. 4. SecureUML+GUI metamodel

where the operation allAssignedRoles() is defined as follows:

context Event::allAssignedRoles():Set(Roles)
body: self.actions.isAssigned.allRoles()$-$>asSet()

Example 7. For any instance of the SecureUML+GUI metamodel that correctly models the graphical user interface in our running example, $EvAu$(onCreateSalary-LabelWindow3) should return Set{Supervisor}. This is because executing the event creating the Salary-label will trigger the data action of reading a *salary* resource, but only supervisors are authorized to execute this data action.

4.1   Smart Security-Aware GUI Models

Informally, a GUI model is smart and security-aware when roles are authorized to execute events *depending on* the actions (both data actions and widget actions) that these events trigger. This dependency relationship, and therefore, the corresponding notion of smartness, can be defined in several ways and we propose one of such definition in this section. Note that our aim is not to give the one "canonical" definition of smartness (should such a definition even exist), but rather to show that non-trivial kinds of smart and security-aware GUI models can be automatically generated from GUI and security-design models using a well-defined model transformation.

To simplify our definition of smartness, in what follows we assume that any model $\mathcal{G}_\mathcal{C}$, with

$$\mathcal{G}_\mathcal{C} = \langle Wd, Wd^c, In, Ev, DaAc, WdAc, WdEv, EvWdAc, EvDaAc \rangle,$$

satisfies the following properties:

1. There are only two widget actions, namely, opening and closing a widget.
2. Every widget has a distinguished event, namely, the event of creating the widget itself. This event has the following properties:
   - If a widget is a non-container widget, then the only widget action associated to this event is the action of opening the widget itself.
   - If a widget is a container widget, then the only actions associated to this event are the actions of opening the widget itself as well as all the widgets that it (immediately) contains.
3. There are no cycles in the widget opening actions. That is, no widget wd can be opened by an action triggered by an event of a widget wd' that was opened by an action triggered by a sequence of events starting from an event of wd.

We denote by open_{wd} (respectively, close_{wd}) the action of opening (respectively, closing) the widget wd. Also, we denote by $EvWdAc^{\circ}(ev)$ the set of opening actions triggered by the event ev, i.e., $EvWdAc^{\circ}(ev) = \{\text{open}_{wd} \mid \text{open}_{wd} \in EvWdAc(ev)\}$. Finally, we denote by onCreate_{wd} the event of creating the widget wd.

Definition 5. *Let C be a component-based model, $C = \langle E, At, As, Md \rangle$. Let $\mathcal{S}_C$ be a security-design model for C,*

$$\mathcal{S}_C = \langle Rs, DaAc, Rl, RsAc, DaAu \rangle,$$

with $Rs = (E \cup At \cup As \cup Md)$. Also, let $\mathcal{G}_C$ be a GUI-model for C,

$$\mathcal{G}_C = \langle Wd, Wd^c, In, Ev, DaAc, WdAc, WdEv, EvWdAc, EvDaAc \rangle.$$

Note that $DaAc$ is shared by $\mathcal{S}_C$ and $\mathcal{G}_C$, i.e., the data actions whose access policy is defined by $\mathcal{S}_C$ are exactly those that can be triggered by events in $\mathcal{G}_C$.

Now, let $\mathcal{M}_{\mathcal{G}_C}$ be a security-GUI model for $\mathcal{G}_C$,

$$\mathcal{M}_{\mathcal{G}_C} = \langle \mathcal{G}_C, Rl, EvAu \rangle.$$

$\mathcal{M}_{\mathcal{G}_C}$ is a smart and security-aware GUI model with respect to $\mathcal{S}_C$ if and only if:

- *The roles that are authorized by $EvAu$ to execute an event ev (different from creating a widget) are exactly those that:*
  - *are also authorized by $DaAu$ to execute all the data actions that will be triggered when executing the event ev, and*
  - *are also authorized by $EvAu$ to create all the widgets that will be opened when executing the event ev (closing a widget, however, is not relevant authorization-wise).*
- *The roles that are authorized by $EvAu$ to create a non-container widget wd are exactly those that are also authorized by $DaAu$ to execute all the data actions that will be triggered when executing this event.*

- *The roles that are authorized by EvAu to create a container widget wd are exactly those that are also authorized by EvAu to create at least one of the widgets (immediately) contained by the widget wd.*

More formally, for any widget $wd \in Wd$ and event $ev \in WdEv(wd)$, the following holds:

- *Case 1: $ev \neq onCreate_{wd}$.*
  Then, $EvAu(ev) =$

$$
= \begin{cases}
\bigcap_{i=1}^{n} DaAu(daac_i) & \text{if } EvDaAc(ev) = \{daac_1, \ldots, daac_n\} \\
& \text{and } EvWdAc^\circ(ev) = \emptyset. \\
\bigcap_{i=1}^{n} EvAu(onCreate_{wd_i}) & \text{if } EvDaAc(ev) = \emptyset \text{ and} \\
& EvWdAc^\circ(ev) = \{open_{wd_1}, \ldots, open_{wd_n}\}. \\
(\bigcap_{i=1}^{n} DaAu(daac_i)) \cap & \\
(\bigcap_{i=1}^{m} EvAu(onCreate_{wd_i})) & \text{if } EvDaAc(ev) = \{daac_1, \ldots, daac_n\} \text{ and} \\
& EvWdAc^\circ(ev) = \{open_{wd_1}, \ldots, open_{wd_m}\}.
\end{cases}
$$

- *Case 2: $ev = onCreate_{wd}$.*
  Then, $EvAu(ev) =$

$$
= \begin{cases}
Rl & \text{if } wd \notin Wd^c \\
& \text{and } EvDaAc(ev) = \emptyset \\
\bigcap_{i=1}^{n} DaAu(daac_i) & \text{if } wd \notin Wd^c \\
& \text{and } EvDaAc(ev) = \{daac_1, \ldots, daac_n\}. \\
\bigcup_{i=1}^{n} EvAu(onCreate_{wd_i}) & \text{if } wd \in Wd^c \\
& \text{and } In(wd) = \{wd_1, \ldots, wd_n\}.
\end{cases}
$$

Lemma 1. *Let C be a component-based model, $C = \langle E, At, As, Md \rangle$. Let S_C be a security-design model for C,*

$$S_C = \langle Rs, DaAc, Rl, RsAc, DaAu \rangle,$$

with $Rs = (E \cup At \cup As \cup Md)$. Also, let $\mathcal{G}_C$ be a GUI-model for C,

$$\mathcal{G}_C = \langle Wd, Wd^c, In, Ev, DaAc, WdAc, WdEv, EvWdAc, EvDaAc \rangle.$$

Then, there exists a unique security-GUI model for $\mathcal{G}_C$ that is smart and security-aware with respect to S_C. This model is denoted by $\mathcal{M}_{(\mathcal{G}_C, S_C)}$,

Proof. Due to space limitations, we only sketch the proof of this lemma here. The crucial point is that, given a security-GUI model $\mathcal{M}_{\mathcal{G}_C} = \langle \mathcal{G}_C, Rl, EvAu \rangle$, the clauses in Definition 5 precisely define, for any event supported by $\mathcal{G}_C$, the set of roles in Rl that should be returned by $EvAu$, if $\mathcal{M}_{\mathcal{G}_C}$ is to be considered smart and security-aware with respect to S_C. First, the clauses in Definition 5 cover all the possible cases. In fact, for every event ev, there is a clause (and only one) that applies to this event. Then, every clause either identifies a specific set of roles as the expected result for $EvAu$ or it recursively calls $EvAu$ on some creating events (namely, those associated to the widgets that will be open when executing the event). Since our GUI models are finite and neither include cycles in the widget opening action nor in the containment-relationship, these recursive calls always terminate. Consequently, for any $\mathcal{S}_C$ and $\mathcal{G}_C$, there always exists a security-GUI model for $\mathcal{G}_C$, namely $\langle \mathcal{G}_C, Rl, EvAuSmart \rangle$, that is smart and security-aware with respect to $\mathcal{S}_C$, where $EvAuSmart : Ev \longrightarrow \mathcal{P}(Rl)$ is the function defined by the clauses in Definition 5 for the given $\mathcal{S}_C$ and $\mathcal{G}_C$.

5   Automatically Generating Smart, Security-GUI Models

In this section, we sketch the definition of a QVT operational transformation `smartandsecure()` that, given a SecureUML+ComponentUML model and a GUI model, automatically generates a SecureUML+GUI model that is both smart and security-aware. The crucial step in this transformation is, of course, the creation of the Permission-objects, and how they link Role-objects to Atomic-Execute-objects (and, through them, to Event-objects). Recall that for the generated model to be considered smart and security-aware, for any of its Event-objects, the value returned by the operation allAssignedRoles()−>asSet(), which "navigates" through those links, must satisfy Definition 5.

We split `smartandsecure()` into two sequential auxiliary model transformations: `generatemodel()` and `addpermissions()`. The method `generatemodel()` generates a SecureUML+GUI model $\mathcal{M}$ that, basically, contains all the widgets, events, widget actions, and data actions, along with their links, that are specified in a given (source) GUI model $\mathcal{G}_C$, plus the roles Rl that are specified in a given (source) SecureUML+ComponentUML $\mathcal{S}_C$. More formally, `generatemodel()` generates a SecureUML+GUI model $\mathcal{M}_{\mathcal{G}_C} = \langle \mathcal{G}_C, Rl, EmptyEvAu \rangle$, where $EmptyEvAu(ev)$ returns, for any event considered in $\mathcal{G}_C$, the empty set of roles, i.e., the expression ev.allAssignedRoles()−>asSet() evaluates to Set{}. The generated model $\mathcal{M}_C$ is not yet security-aware or smart (unless the given security-design model $\mathcal{S}_C$ does not specify any authorization restrictions.)

Next, based on the results of evaluating an OCL operation `EvAuSmart()` (whose definition directly translates into OCL the clauses in Definition 5), the method `addpermission()` augments the SecureUML+GUI model $\mathcal{M}_{\mathcal{G}_C}$ with all the permissions that makes this model smart and security-aware with respect to $\mathcal{S}_C$. More concretely, it adds all the permissions, along with their links to the appropriate roles and atomic execute actions, that are required for the following to hold: for any event ev, the expression ev.actions.isAssigned.allRoles()−>asSet() evaluates to the set of roles that should be authorized to execute the event ev if $\mathcal{M}_{\mathcal{G}_C}$ is to be considered smart and security-aware with respect to $\mathcal{S}_C$.

6   Related and Future Work

Creating user interfaces is a common task in application development and one that is often time consuming and therefore expensive. There have been numerous proposals and tools that aim to reduce the effort required to build effective, user-friendly graphical interfaces. Surprisingly, there has been no prior research until now on the systematic design of GUIs whose functionality should adhere to the security policy of the underlying application-data model. The idea we develop here originated in [12], where we first proposed using model-transformations to generate simple (not necessarily smart) security-aware GUI models.

In the modeling community, other researchers have investigated how to extend existing modeling languages for GUI modeling. [4] proposes a UML profile to model GUI layout. We do not consider layout issues since translating security from data models into GUI models is independent (except for containment hierarchies, which we do consider) of the graphical appearance and location of the widgets. However, we do plan to use widgets' graphical information to make our GUIs more appealing to users. [3] proposes a heavyweight, template-based extension of UML for GUI modeling to help develop GUIs for large-scale systems, although access control decisions are not part of this work either. This work is similar to ours in that it also separates concerns within the modeling phase by separating the construction of the application and the GUIs. Our modeling techniques also differ since we use metamodeling in contrast to their use of stereotypes. In [3], the task of the GUI designer is reduced to choosing between *window types*, which are parameterized templates (including interaction behavior and layout). If we had used stereotyped GUI designs in our approach, we would have provided different fixed GUI designs that could be made automatically smart and security-aware according to the underlying data model security policy.

[6] approaches security as a *crosscutting concern* in terms of aspect-oriented programming for a software system. The main contribution of this work is a behavioral definition of aspects. The authors propose enriching a data model with a security policy by performing a model transformation using the bidirectional object-oriented transformation language (BOTL) [9,10]. Although they present examples, they only outline a method for model transformation based on the proposed definition. Therefore a comparison here would be difficult. Notice also that such comparison would concern our integration of the SecureUML policies in the data model, not the GUI model generation itself. In [13], the authors present a model transformation methodology to integrate non-functional requirements, such as security, in a model-driven software product line. In this setting, abstract design models of the application and its security are built and these models are refined in parallel using model transformation to obtain an implementation model with Java Platform, Enterprise Edition (JEE) security annotations. In contrast, we integrate the data, the GUI, and the security platform-independent models (PIMs) from the design stage to obtain a security-aware GUI PIM from which one can generate a functional GUI, which will provide a security-aware and smart access to the data. Regarding GUI generation, there are a number of tools for the automated design and generation of user interfaces, which can generate the static layout of an interface from the application's data model. However, security concerns are not among the problems addressed.

In the programming community, independent of model-driven initiatives, numerous projects have addressed the problem of how to best implement graphical user interfaces for application data. For example, [8] proposes enriching the application source code with annotations that control the generation of the graphical user interfaces. Other researchers have designed and implemented specialized tools that generate graphical user interfaces meeting their own specific requirements. These tools simplify configuring personal services, enabling the combination of different kinds of events [11]. Also, there are many GUI builders, either integrated into IDEs or available as plug-ins, that simplify the task of creating application GUIs in different programming languages. However, to the best of our knowledge, [7] is the only tool capable (although still a prototype) of automatically generating smart, security-aware functional GUIs.

Finally, our work is also related to research in the field of intelligent user interfaces. In our view, smart security-aware GUIs can be seen as a class of intelligent user interfaces. Our GUIs take advantage of the users' status to tailor their access to the application data. An interesting follow up question concerns the generality of our model-transformation approach and whether it can be used to generate other classes of intelligent interfaces.

7   Conclusions

We have presented an approach based on model-transformation for automatically generating smart, security-aware GUIs. Given an application-data model and a GUI model, our transformation makes the GUI model both smart and security-aware. We have implemented our approach using the Operational QVT transformation engine that is provided within Eclipse.

As a design methodology, our approach has three main advantages over traditional approaches to software design. First, security engineers and GUI designers can independently model what they know best. Second, security engineers and GUI designers can independently change their models, and these changes are automatically propagated to the security-aware GUI models. Third, GUI designers can use the generated security-aware GUI models to check that they are designing the right GUI to give the (authorized) users access to the (intended) application data.

The work presented here is the corner stone of a more ambitious project for making model-driven security an effective and useful approach for generating multiple layers of security-critical systems in industrial software development.

References

1. Basin, D., Clavel, M., Doser, J., Egea, M.: Automated analysis of security-design models. Information and Software Technology 51(5), 815–831 (2009)
2. Basin, D., Doser, J., Lodderstedt, T.: Model driven security: From UML models to access control infrastructures. ACM Transactions on Software Engineering and Methodology 15(1), 39–91 (2006)

3. Blankenhorn, K., Walter, W.: Extending UML to GUI modeling (2004),
   http://www.bitfolge.de/pubs/MC2004_Poster_Blankenhorn.pdf
4. TATA Research Development and Design Center. Heavyweight extension of UML
   for GUI modeling: A template based approach (2001),
   http://www.omg.org/news/meetings/workshops/presentations/
   uml2001_presentations/10-2_Venkatesh_typesasStereotypes.pdf
5. Ferraiolo, D.F., Sandhu, R.S., Gavrila, S., Kuhn, D.R., Chandramouli, R.: Proposed NIST standard for Role-Based access control. ACM Transactions on Information and System Security 4(3), 224–274 (2001)
6. Fox, J., Jürjens, J.: Introducing security aspects with model transformations. In:
   12th IEEE International Conference on the Engineering of Computer-Based Systems (ECBS 2005), Greenbelt, MD, USA, April 4-7, pp. 543–549 (2005)
7. BM1 Software Group. The SmartGUI Project (2009),
   http://www.bm1software.com/
8. Jelinek, J., Slavik, P.: GUI generation from annotated source code. In: TAMODIA
   2004: Proceedings of the 3rd annual Conference on Task Models and Diagrams,
   pp. 129–136. ACM, New York (2004)
9. Marschall, F., Braun, P.: Model transformations for the MDA with BOTL. Technical report, University of Twente (2003)
10. Marschall, F., Braun, P.: Bidirectional object oriented transformation language
    (2005), http://sourceforge.net/projects/botl/
11. Ogura, M., Mineno, H., Ishikaw, N., Osano, T., Mizuno, T.: Automatic GUI Gen-
    eration for Meta-data Based PUCC Sensor Gateway. In: Lovrek, I., Howlett, R.J.,
    Jain, L.C. (eds.) KES 2008, Part III. LNCS (LNAI), vol. 5179, pp. 159–166.
    Springer, Heidelberg (2008)
12. Schläpfer, M., Egea, M., Basin, D., Clavel, M.: Automatic generation of security-
    aware GUI models. In: Bagnato, A. (ed.) European Workshop on Security in
    Model Driven Arquitecture 2009 (SEC-MDA 2009). Workshop Proceedings Series,
    vol. WP09-06, pp. 42–56. CTIT, Enschede (2009)
13. Yie, A., Casallas, R., Deridder, D., Van Der Straeten, R.: Multi-step concern re-
    finement. In: EA 2008: Proceedings of the 2008 AOSD workshop on Early aspects,
    pp. 1–8. ACM, New York (2008)

Report: Modular Safeguards to Create Holistic Security Requirement Specifications for System of Systems

Albin Zuccato[1,2], Nils Daniels[1], Cheevarat Jampathom[1], and Mikael Nilson[1]

[1] TeliaSonera, Common Development, Product Security, Sweden
[2] Stockholm University, Department of Computer and System Science, Sweden

Abstract. The specification of security requirements for systems of systems is often an activity that is forced upon non-security experts and performed under time pressure. This paper describes how we have addressed this problem by using a collection of modular safeguards, which are tailored to the application domain. These safeguards, which are specific but still fairly atomic, are combined into requirement profiles that seamlessly integrate into the overall development approach. These safeguards are grouped into 15 classes which subsume requirements that aim for low, medium and high security capabilities. Each requirement is further specified with a technical description defining actual values. To achieve a holistic coverage, we have created requirement profiles that define combinations of modular safeguards and have added complementary organizational safeguards. We will show how we have developed this approach over the years and present our practical experiences of the seamless integration into the development life cycle.

1 Introduction

Security requirements, as a special type of general system requirements, are an established way to express what kind of security is expected from a (telecommunication) service or product – which are a system of systems. Those security requirements are an important cornerstone for a secure offering that satisfies market demands. Within literature we find an almost unanimous consensus that security requirements shall be elucidated at the beginning of the development and are supposed to guide all future security activities like implementation, assurance, operation and retirement. However, although everybody agrees that security requirements are needed it seems quite common that they come as an "$\epsilon \nu \rho \eta \kappa \alpha$" (heureka) experience to the security expert (given more or less extensive preparation by means of risk analysis or similar). Sometimes the security expert is assisted by a more or less elaborate process. But ultimately it is their "creativity" that leads to the security requirements.

From a practical viewpoint, we see a number of issues that are difficult to satisfy with such thinking. First of all it is assumed that there is a security expert involved. Although desirable, the practical scarcity of security expertise

F. Massacci, D. Wallach, and N. Zannone (Eds.): ESSoS 2010, LNCS 5965, pp. 218–230, 2010.

implies that only security critical projects receive this level of attention. And even when a security expert at hand, it is likely that this person is under time critical restrains with little time available to produce solid requirement specifications. Secondly, it is taken for granted that the resulting security requirements are detailed enough and unambiguous, but also simple enough to be directly understandable to developers. Furthermore, we have observed wishes that security requirements "per se" be uniformed enough to foster generic and reusable security solutions. From our own practical experiences and through investigations of existing approaches we find that this is hardly the case.

We therefore developed an approach that takes a set of modular security safeguards, which are specific and fairly atomic, and assembles them into holistic security requirement profiles. These profiles take the interdependencies and synergies between the modular safeguards into account and complements them with higher level organizational security mechanisms. Such profiles can be used subsequently to an information classification [1]. We will show the advantages of this approach on of our projects. We then describe the practical experiences we made and indicate a number of advantages we have seen. Finally we deliver our conclusions and suggest a way forward.

2   Security Requirement Engineering Today

To describe good security requirements for information systems is sometimes considered in between engineering and art[2]. This is due to the significant portion of creativity that is commonly assumed to be required. Although this might be true to some extent, it is desirable to encapsulate the creativity into well defined areas and have a structured engineering process at hand. Subsequently, we present a selection of approaches that can be considered when tasked to find the (security) requirements for a system and show some practical issues for them.

We would like to start with Common Criteria (CC)[3] which we consider the most important approach in this area. CC offers a modular set of requirements with a very wide coverage. They are well written and comprehensible. Given enough time and expertise, CC would be very suitable for the system of systems security area. Unfortunately, both time and expertise must be considered scarce in normal industrial development. The biggest challenge with CC is how to combine the safeguards in a suitable way. It is not clearly indicated which safeguards should be considered indispensable, which are situation dependent and which are "nice to have". Furthermore, significant security architecture skills are required to solve the interdependencies between classes as CC only provides a dependency indication inside the classes. If a protection profile (a CC construct for requirements for an explicit and well defined purpose) is available this would be solved. However, to derive a protection profile implies a significant work effort which, from an industrial point of view, seldom is feasible. Another problem we see is that although CC indicates the class, it is the author of a security target or protection profile (i.e. the requirement specification) that has to define the actual values for each requirement. We also found that CC has no clear interface towards earlier security demand assessments (e.g. risk analysis). It requires expert knowledge to figure out which component(s) mitigate which risk or satisfy which demand. This issue is hardly addressed anywhere in CC. Altogether this requires a security specialist whom is highly familiar with CC – something that is not commonly available for the average system engineering project. We therefore conclude that although CC is technically very sophisticated there are "soft" issues such as an extensive need for security expertise and high application threshold due to size and educational effort as well as significant overhead that makes the usage for Common Criteria too problematic in a practical day-to-day system development.

Another interesting group is the "monolithic approaches" which provides full security requirement sets for specific areas. The most prominent among them which we consider are the specifications published by NIST in their SP800 series[4]. These special publications are organized within areas and usually provide a very good coverage of the topic. However, they are extremely monolithic and a normal system of systems covers many areas addressed in various SP's. This leads to significant redundancy and makes it difficult to apply the holistic view necessary for system security engineering. We have found that industrial development activities suffer under significant time pressure and consider it therefore too optimistic to assume that projects, that have to deal with multiple systems and multiple security requirements, would have the time and knowledge to harmonize requirements to derive a non-redundant and holistic set. Furthermore, we find that the threshold to access them is quite high due to the fact that each SP is around 50 to 100 pages and one would be required to read several of them. Other approaches in this group are the IEEE RFC's or PCI DSS. They suffer from the same problems of redundancy, application threshold and monolithic thinking.

As a third group of approaches we found "'security requirement engineering methods". Among them we find approaches like HSRE [5] or SQUARE [6]. These approaches investigate various sources and suggest different elicitation techniques to find security requirements. Unfortunately, to deliver the excellent requirement specifications they promise a security expert, who select and performs the suitable activities, is required. We also find that these methodologies describe the "what" but not the "how". This implies for us that a high adaptation and education threshold must be overcome.

A fourth group of approaches we would like to describe as "methods and best practices". A typical example for such approaches are the use-case based abuse-cases [7] and misuse cases [8]. Both aim to apply standard requirement engineering techniques to security requirements. Another example is ISO27001[9] which contains instructions of how to derive requirements from a list of requirement areas. However, they have too high abstraction level and are not sufficiently detailed to serve as measurable requirements.

Although this list is not exhaustive, we cannot present more due to space constrains. In general, we are confident to say that these approaches commonly provide a solid structure, process and some good advice of how to derive security requirements. However, they usually neither provide ready-made requirement sets nor implementation guidelines and therefore have a fairly high threshold before they can become applicable. We also find that a great deal of security expertise is required in order to apply these approaches.

3   Security Requirement Specification Engineering for System of Systems

Telecommunication products and services are often extremely large and complex systems made up from other systems. These systems of systems have both overall and specific requirements for each subsystem. When constructing such systems of systems it is more common to integrate existing systems rather than to develop new ones. This implies that some of the sub-systems cannot be modified too much as it might affect the systems functionality. The requirement specification process therefore must take those limitations into account.

From an industrial perspective the requirement specification is further constrained by factors like execution time, lead time and budget restrictions. However, most significant is the lack of security expertise for all projects. Security expert resources are scarce and have to focus on activities where they can be of most benefit. This implies that the average development projects receives very little expert attention and is left to its own accord. To derive security requirements under these conditions, it is necessary to have a process that requires as little security knowledge as possible. Furthermore, such a process must be perceived as part of the daily work and needs therefore to have tight integration into the normal development life cycle.

To accommodate with the complexity, the lack of resources and the need for compliance with regulations and policies, we have developed a requirement specification approach that has very little freedom but performs well for non-security-experts. The foundation of this approach is a collection of modular safeguards. These safeguards are well encapsulated and have clear interfaces. This implies that they are tailored to the application domain to be sufficiently specific and that they are fairly atomic. The safeguards are grouped into approximately 15 classes, which subsume requirements that aim for no, low, medium and high security capabilities. Each requirement is further specified with a technical description defining actual values. In order to achieve a holistic coverage, we have created requirement profiles that define combinations of necessary modular safeguards as well as complementary organizational safeguards. From a usage perspective they only have to pick the appropriate profile for the security demand as defined by the information classification. This will automatically yield all necessary safeguards and their associated technical specifications, i.e. the security requirements.

3.1   Modular Security Requirement Classes

Over the years there have been constant discussions in our organization on whether different security mechanisms (safeguards) were strong enough or not.

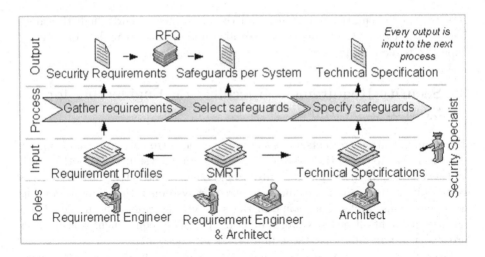

Fig. 1. Requirement Engineering Process

We think that this should depend on the "relative" strength needed and not on an abstract "absolute" strength. It is therefore necessary to express the protection demand in a hierarchical way and to have corresponding levels of security safeguards. The first part with the demands are realized with the consequence reference table we find in [1]. The corresponding Security Mechanism Reference Table (SMRT) is but a logical continuation to describe which requirements that satisfy each demand. From an organizational policy point of view, a desirable prerequisite for SMRT is that it have to be fairly stable over time. Thus the security safeguards need to be described in a functional (rather then technical) and modular manner. The goal is that SMRT serves as a repository for requirements and can be used to evaluate the security protection levels of existing solutions. This way we save development and assessment time and make security accessible for the average development engineer.

We started by looking at Common Criteria (CC) [10] to look for safeguards to choose from. Unfortunately we found CC in its entirety to be too complex for our needs. Especially the unsuitability for non-security experts was a preventive argument. Instead we browsed CC for security mechanisms that suited our application domain and intention. For each security mechanism, we defined the protection level it could offer (valid in our organization) for each of the security properties confidentiality, integrity and availability. For levels 0 – 3 we define as none, low, moderate and high security.

This initial list from CC was complemented with safeguards that were used within our organization. This yielded a quite extensive list of safeguards which we needed to reduce to accommodate economic needs and minimize usage complexity. The goal was to allow simpler usage and achieve easier comprehension for the non-experts. This led us to a list containing about 3 safeguard groups with a total of approximately 15 categories, see Table 1, describing security mechanisms for level zero to three – see Table 2.

Table 1. Security mechanism reference table

System dependent	User authentication	Intrusion detection and response
	Access control	Disaster recovery
	Stored data integrity	...
	Stored data confidentiality	
Interdependent	Traceability	Restorability
	Key management	...
Network dependent	Data integrity transfer protection	Network Authentication
	Data confidentiality transfer protection	Pervasive

Table 2. Example for functional and modular authentication mechanism

Lvl	Name	Description
0	One-factor, clear	secret (e.g. password) is stored or transmitted in clear text
1	One-factor, protected	secret (e.g. password) is not stored or transmitted in clear text
2	One-factor, strong	Lvl 1 + construction requirements, locking mechanisms and issuing guidelines
2	Two-factor, non-repeatable	Radius-based in order to centralize the authentication method
2	Biometrics	Biometric authentication mechanisms were FRA and FRR meet at no more then 3%
3	Two-factor, physical	As level 2 plus one factor is a physical token (e.g. calculator, smart card, biometrics ...).

Concerning the usage scenario we formed a two-step process. First, we introduced a baseline security level in which the indispensable security safeguards were integrated. This baseline is mandatory for all projects and serves primarily the purpose of infrastructure protection. Secondly, every security development and maintenance activity is to consider the required protection level and add security mechanisms to that extent. Concerning the possible combinations we initially started out with a matrix indicating which mechanisms to combine. This turned out to be too complicated in the long run as much work effort had to be spent by each development project to create the requirement sets by them selves. We therefore improved the approach as presented within the next chapter.

3.2   Security Requirement Profiles

Following up on projects that have used our approach we learned that it was extremely difficult for non-security-experts to achieve a holistic security coverage that satisfied the corporate security policies. As a reason we found that product developers had difficulties in knowing which combination of safeguards that where necessary for the required security protection. Especially when constrains of cost effectiveness and non-interference with functionality and user experience had to be considered. Our solution was to create security profiles which translate policies and other steering documents into practical and reasonable security requirements that are understandable and applicable. Due to the fact that we use information security classification[1] we found that the majority of projects have their security demands expressed in four levels. For these four levels we crafted three requirement profiles (for "none" we do not have a profile as the baseline covers it sufficiently).

We started out by creating suitable combinations of the security safeguards from the security mechanism reference table. As the reference table is fairly extensive we had various mechanisms at each level which had different interdependencies and we needed to select the desirable alternatives. An example is the confidentiality protection of information in a system. We could then choose between stored data confidentiality safeguards, which are different types of cryptography, that need to be combined with key management or access control mechanisms combined with authentication and audit trails. Although the overall security was important we also had to consider other factors like performance impacts, implementation costs, maintenance efforts and operation costs. The derived draft profiles where a great improvement. However, during the prototyping process we found that we could integrate organizational safeguards to fill eventual gaps between mechanisms and assure a full holistic coverage and the intended policy compliance.

To be able to reproduce this we documented our findings in an algorithm. The first step was that only mechanisms at the same level were allowed in a profile. The second step was to deal with the interdependencies. For this we created a matrix that indicated the correlation between mechanisms as synergies, dependencies, conflicts and redundancies. When we constructed a profile, mechanisms that had synergies were primarily chosen. When it came to dependencies, we aimed to reduce them as much as possible and documented the motivation for why and how we reduced the profiles. For the remaining two, our strategy was to avoid conflicts at all possible costs by choosing alternative mechanisms and by eliminating redundancies when we did not have dependencies. The third step was to identify eventual gaps between mechanisms and described organizational safeguards to mitigate them.

While creating the profiles and performing all trade-offs we could see the difficulties that non-security-experts had to deal with. We found that a great deal of expertise and pragmatism is necessary and that the profiles could never be perfect. However, we have achieved a reasonable balance that is sufficiently accurate to function without modification in our average projects. For the extraordinary projects, the profiles provide a solid base for the security expert to perform the fine tuning and to accommodate to their specific needs.

3.3 Technical Specification

The above presented Security Mechanism Reference Table (SMRT) and profiles were a very good starting point. We used SMRT for more then seven years and found that it required a bit too much security expertise to use it to its full potential for the technical specification part and the subsequent design. SMRT and our profiles indicated what we wanted and how they could be combined. The profiles failed, however, to communicate clearly what security mechanisms we had in mind. Security specialists frequently received the questions from developers and suppliers what it actually meant. It was necessary to provide interpretations on a case-by-case base which turned out to consume some extra time and resources. We also realized that the chosen solutions had significant variations. This was not desirable as it prevented economy of scales where we could use the same solution for various products and by those means to reduce the development and operational costs.

To solve this problem we started to investigate the common solutions that were derived in the different projects and created a condensed set of refined technical specification requirements. We verified these technical specifications against existing suggestions in literature. For example Common Criteria [3] or the security patterns community[11]. A typical technical specification would define all the security mechanisms for a requirement at a given protection level and provide actual values for the parameters. An example for password to be used on SMRT level 2 can be seen below. This example was derived from our demand and internal usage of strong passwords, security patterns [11] for authentication. To derive and verify the strength we used the algorithms suggested in NIST SP800-63[12].

```
Password, to comply with one-factor, strong requirement

Length: 6-8 characters
Composition: ASCII 32-126, 129-165, at least one lower case, upper case and number
             (allow for special character but not required)
Source:    Individual (recommendations provided) or automatic (which is stronger)
Lifetime:  min 1 day, recommended 3 month, max. 6 month
Ownership: Individual (should be made clear in the contract
Authentication period: log-in and after 15 minutes of terminal inactivity
Authentication exception: 3 failed login lock the account; reset is possible via one-time
             password by SMS (then for only 15 minutes) or paper mail to enter new password
Distribution: via paper mail a random PW is distributed, must be changed at first login
Storage:   Only hashed with a cryptographic Hash (preferably SHA-256)
Transmission: Only encrypted (preferably replay-protected by seeded hash before encrypted)
```

This step was highly appreciated by the developers and system architects as they could now derive their security requirements with associated technical specifications without having to wait for the security experts to refine the requirements. We found that both the effort and the lead time for security requirement engineering were significantly reduced. Secondary effects like reuse and complexity reduction are harder to measure, especially given the short time of usage of about one year, but we think that we see clear indications of these positive effects.

3.4   General Applicability

In the introduction we mentioned that the approach is fairly tailored to the application domain. We are, however, convinced that the structure and artifacts are generic enough to be useable in other domains. We also find that information classification becomes more and more used[1]. Our approach is a logical continuation of that work and we have received external confirmation that it can be valuable. We therefore investigate all three components and present our ideas on how to apply them outside of our scope.

---

[1] At workshops and personal interviews with the Swedish National Emergency Response Authority, Swedish Banks and the Swedish National Pharmacy Corporation.

When it comes to the security mechanism reference table the modular structure per se assures that it can be adapted to any domain. It might not even be necessary to craft an own SMRT but use the categories and classes we introduce. Eventually the values should be verified to match ones own risk appetite. We also think that defining complementary classes and mechanisms for the own application domain is fairly easy due to the modular structure.

The requirement specification profiles are based on our risk appetite and infrastructure. To apply them would require the same preconditions. However, the general algorithm, especially the interdependency matrix, should be reusable. It is again possible to reassess the dependencies for ones own infrastructure and risk appetite if it seems unsatisfactory.

Finally the technical specification is the least generic one. It is heavily tailored to our environment and the values clearly reflect our risk appetite. What we think can be reused with advantage are the structures and patterns for each mechanism and the verification infrastructure we have associated to some of them.

4   Example

This section presents an example of how to apply holistic security requirement profiles to a Mobile Push Email (MPE) service. MPE provides an always-on functionality where the users are able to access their emails anywhere anytime. The idea is to support our customers demand for mobility. To enable this service the operator needs the following extension components to the existing mail infrastructure:

- Relay Server - is the central server responsible for message routing which enables the mobile client to communicate with push connector, and forward email messages to the end-user.
- Push connector - is the application which connects to an email server, and transmits email content to the mobile client.
- Support site - is the online fault reporting system.

The first step, in order to retrieve a correct level of security requirement profile, is to list the security-relevant information handled in MPE and classify it – see [1]. We focus only on the components that are new to the service – see Table 3. The Relay server and connector handle the user's email account credentials and email contents. These credential information is then used to authenticate the user against the back-end email server. This information is "access control data" with sensitivity level 3. Once the user is authenticated, the user's emails are pushed (actively transferred) to the mobile client. The email content is considered as "user traffic" with sensitivity level 2. The support site handles log files and possibly some customer related information. This information types ("logs" and "user information") is classified as level 2.

After classifying the information we define the maximum protection level needed (for each system). We assume that the security of a system is only as strong as its weakest part and therefore use a strategy in which the whole system

Table 3. Information Classification for Mobile Push Email

System	Information	Description	Information type	Protection demand
Relay Server & Connector	Email contents	Email that are pushed to mobile	User traffic (Lvl 2)	High (Lvl 3)
	Email account credentials	Login information for user authentication to mail server	AC data (Lvl 3)	
Support site	Log files	Log files that describe problems	Logs (Lvl 2)	Moderate (Lvl 2)
	Support ticket data	Detailed customer data might be in the log files	User info. (Lvl 2)	

protection has to satisfy its maximum demand level. If, for example a system handles at least one set of level 3 information, that system automatically will require high security protection (lvl 3). In our example the relay server and connector requires high (lvl 3) protection whereas the support system requires moderate (lvl 2) protection. This implies that for the support system, the "moderate" profile – see Table 4 – has to be used.

Table 4. Moderate security requirement profile (level 2)

Requirement Areas	Requirement Name	Requirement Description
Authentication	One factor, strong	secret (e.g. password) is not stored or transmitted in clear text; construction requirements, locking mechanisms and issuing guidelines.
Access control	Discretionary access control	Access rights granted by the object's owner. The access rights have to be reviewed regularly in long interval (e.g. one year OR when changed).
Data confidentiality transfer protection ...	Strong encryption of a communication path	Strong encryption (e.g. 3DES, AES) shall be used. Weak encryption algorithms (e.g. DES) are not allowed.

5   Experiences

The above presented processes and documents have been iteratively developed over the last nine years. We were inspire by the ideas implied in ISO9000[13] and SSE-CMM[14] and improved the method based on measuring real world projects. As the method is part of the mainstream development life cycle, it is performed by approximately 200 projects per year. One important lesson we learned is that cost efficiency and complexity reduction are key factors for success. This implies that "good enough" security which is implemented and used is ultimately better then optimal security on paper. Subsequently, we shall go through the positive and negative experiences we had in more detail.

Most importantly for us was that the developers considered that the security requirement engineering became more understandable to them and that the usability of the method was improved. The need for expert involvement decreased significantly and since the introduction of profiles most average projects consider it possible to perform the security requirement specification on their own. Noteworthy is also that due to the iterative improvements the developers felt integrated and developed a positive attitude.

In respect to cost efficiency and willingness to perform security work we found that the time effort was of big significance. We therefore investigated how much time was required for each project and actively worked with time issues – see Figure 2. Specifically on the execution time we found a five-fold reduction over the years. This in turn allowed projects to considerably reduce the amount of man-hours needed in developing requirement specifications which could then be forward to suppliers in shorter delivery time frames. Also important in the light of shorter time-to-market cycles was a three-fold reduction in lead time. Tasks that needed to wait on one another were then eliminated by having many activities perform "semi-automatic" due to the clear guidelines available. We also found that lead time was reduced as the security experts could then prioritize focus to critical tasks instead of covering a multitude of tasks.

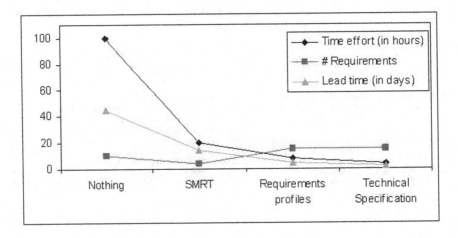

Fig. 2. Experiences of security requirement profiles

When it comes to the quality of requirements we found that in the beginning we had less requirements which had, however, a better coverage. Over time we also increased the amount of requirements by maintaining the holistic coverage. As most of the requirements are now pre-fabricated the ambiguity is reduced. An important positive side effect is that it is far easier now to show policy compliance for all projects without the projects to worry about this issue. And last but not least, the redundancy of requirements has reduced which has positive economic effects on the development and operational costs.

When it comes to the limitations of this method, it is obvious that this method applied to its full extent has hardly any freedom of choice. The profiles are not fully optimized but rather focus to catch as many cases as possible. We consider it possible to perform an optimization of the result by further tailoring on a case-by-case bases which requires security expert knowledge. We therefore think that a security expert can use the method for guidance but should not feel constrained to enhance where reasonable.

We also have to emphasize that the strength description of safeguards is fairly subjective and heavily guided by the risk appetite. We found that we could establish an agreement concerning the strength inside our organization but are well aware that this is not necessarily generic. Although subjective we found that we have something to discuss and compare when talking to third parties and establish at least a clear picture of different perceptions.

Finally we found that quality assurance and verification is an issue. Mismatches between profiles and a project's requirement specification that are detected in assurance activities require active treatment. Especially when shortcomings are perceived a special exception handling process is necessary. During such an exception handling a risk analysis is commonly performed. On the positive side, we found that such an analysis is fortunately limited in scope and has to investigate only the problem cases. However, expert resources for this activity have to be taken into account that do not concern the individual projects but have an impact on the overall performance.

6   Conclusion

As a telecom provider we have a need for security in our infrastructure and services. As this infrastructure, and the services running on it, is mainly systems of systems we need an approach that can provide us with multidimensional and holistic security requirements. Current approaches require security expertise which makes them difficult to apply in an industrial setting. We therefore present an approach that builds on a flexible and extensible modular safeguard base which is narrowed down into directly applicable security requirement profiles to remove uncertainty and ambiguity (but also the freedom) for system developers. As an extension we also showed the capability to attach technical specifications to further reduce the need for security expertise.

In the experience section we have presented our positive and negative experiences with this approach. We would like to summarize these experiences by claiming that this approach is simple to use out of the box (like monolithic approaches) and extensible when necessary, due to its modular base (like the CC) and the well defined construction algorithms are suitable for adaptation and optimization (like the best practise approaches).

For future development our focus currently lies on crafting further technical specifications and keeping our mechanism library up-to-date. Previously we have also described a testing approach that allows the non-expert to find appropriate tests[15]. This approach must be further integrated into the profiles to automatically derive the required test specifications and test cases. In addition we aim to further collect empirical data about the process efficiency to be able to identify areas were further improvement (e.g. automation, tool development, etc.) is possible.

References

1. Zuccato, A., Endersz, V., Daniels, N.: Security requirement Engineering at a Telekom Provider. In: Jakoubi, S., Tjoa, S., Weippl, E. (eds.) ARES 2008 Proceedings, pp. 1139–1147. IEEE Computer Society, Los Alamitos (2008)
2. Bishop, M.: Computer Security: Art and Science. Addison Wesley, Reading (2003)
3. International Organization for Standardization: ISO/IEC 15408:2005 - Common Criteria for Information Technology Evaluation (2005)
4. National Institute of Standards and Technology: Special publications (800 series) (2009), http://csrc.nist.gov/publications/PubsSPs.html
5. Zuccato, A.: Holistic security requirement engineering for electronic commerce. Computers & Security 23(1), 63–76 (2004)
6. Mead, N., Hough, E., Stehney II, T.: Security Quality Requirements Engineering (SQUARE) Methodology. SEI Technical Report (2005)
7. McDermott, J., Fox, C.: Using abuse case models for security requirements analysis. In: Proceeding of the 15th Annual Computer Security Applications Conference, pp. 55–64. IEEE, Los Alamitos (1999)
8. Sindre, G., Opdahl, A.L.: Eliciting security requirements with misuse cases. Requirements Engineering 10(1), 34–44 (2005)
9. International Organization for Standardization: ISO/IEC 27001:2005, IInformation technology – Security techniques – Information security management systems – Requirements (2005)
10. International Organization for Standardization: ISO/IEC 15408-2:1999 Information technology – Security techniques – Evaluation criteria for IT security – Part 2: Security functional requirements (1999)
11. Schumacher, M., Fernandez-Buglioni, E., abd Frank Buschman, D.H., Sommerlad, P.: Security Patterns - Integrating Security and Systems Engineering. Wiley, Chichester (2006)
12. Burr, W.E., Dodson, D.F., Polk, W.T.: Electronic Authentication Guideline. NIST Special Publication 800-63 Version 1.0.2, National Institute of Standards and Technology (2006)
13. International Organization for Standardization: ISO/IEC 9000:2000 Quality management systems - Fundamentals and vocabulary (2000)
14. SSE-CMM Project: Systems Security Engineering Capability Maturity Model. v 3.0 edn. (2003)
15. Zuccato, A., Kögler, C.: Functional security testing – closing the gap between software testing and security testing: A case from a telecom provider. In: Massacci, F., Redwine Jr., S.T., Zannone, N. (eds.) ESSoS 2009. LNCS, vol. 5429, pp. 185–194. Springer, Heidelberg (2009)

Idea: A Feasibility Study in Model Based Prediction of Impact of Changes on System Quality

Aida Omerovic[1,2], Anette Andresen[3], Håvard Grindheim[3], Per Myrseth[3], Atle Refsdal[1], Ketil Stølen[1,2], and Jon Ølnes[3]

[1] SINTEF ICT, Norway
[2] University of Oslo, Norway
[3] DNV, Norway

Abstract. We propose a method, called PREDIQT, for model based prediction of impact of architecture design changes on system quality. PREDIQT supports simultaneous analysis of several quality attributes and their trade-offs. This paper argues for the feasibility of the PREDIQT method based on a comprehensive industrial case study targeting a system for managing validation of electronic certificates and signatures worldwide. We give an overview of the PREDIQT method, and present an evaluation of the method in terms of a feasibility study.

1 Introduction

When adapting a system to new usage patterns or technologies, it is necessary to foresee what such adaptions of architectural design imply in terms of system quality. Predictability with respect to non-functional requirements is one of the necessary conditions for the trustworthiness of a system.

We have developed the PREDIQT method with the aim to facilitate prediction of impacts of architecture design changes on system quality. The PREDIQT method produces and applies a multi-layer model structure, called prediction models, which represent system design, system quality and the interrelationship between the two. Our overall hypothesis is that the PREDIQT method can, within practical settings and with needed accuracy, be used to predict the effect of specified architecture design changes, on the quality of a system. Quality is decomposed through a set of quality attributes, relevant for the target system. PREDIQT supports simultaneous analysis of all the identified quality attributes and their trade-offs. The PREDIQT approach of merging quality concepts and design models into multiple, quality attribute oriented "Dependency Views" (DVs), and thereafter simulating impacts of architecture design changes on quality, is novel.

This paper reports on experiences from using the PREDIQT method in a major industrial case study focusing on a so-called "Validation Authority" (VA) system [10] for evaluation of electronic identifiers worldwide. We first give an overview of the PREDIQT method, and then present an evaluation of the method

F. Massacci, D. Wallach, and N. Zannone (Eds.): ESSoS 2010, LNCS 5965, pp. 231–240, 2010.

in terms of a feasibility study. The evaluation focuses on security and its trade-offs with the overall quality attributes identified and defined with respect to the VA system, namely scalability and availability. The results indicate that PREDIQT is feasible in practice, in the sense that it can be carried out on a realistic industrial case and with limited effort and resources.

The paper is organized as follows: An overview of the PREDIQT method is provided in Section 2. Section 3 presents the feasibility study. Section 4 provides an overview of related research. Concluding remarks are summarized in Section 5. See the full technical report [12] for a more detailed presentation.

2   Overview of the the PREDIQT Method

The process of the PREDIQT method consists of the three overall phases illustrated by Figure 1. Each of these phases is decomposed into sub-phases. Sections 2.1 and 2.2 present the "Target modeling" and "Verification of prediction models" phases, respectively. The "Application of prediction models" phase consists of the sub-phases presented in Section 2.3.

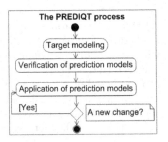

Fig. 1. The overall PREDIQT process

2.1   Target Modeling

The sub-phases within the "Target modeling" phase are depicted in Figure 2. The requirement specifications and system design models are assumed to be made available to the analysis team, along with the intended usage, operational environment constraints (if available) and expected nature and rate of changes.

Characterize the target and the objectives: Based on the initial input, the stakeholders involved deduce a high level characterization of the target system, its scope and the objectives of the prediction analysis, by formulating the system boundaries, system context (including the operational profile), system life time and the extent (nature and rate) of design changes expected.

Create quality models: Quality models are created in the form of a tree, by decomposing total quality into the system specific quality attributes and their respective sub-characteristics. The quality models represent a taxonomy with interpretations and formal definitions of system quality notions.

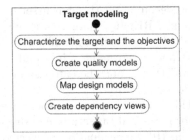

Fig. 2. Target modeling phase

Map design models: The initially obtained design models are customized so that (1) only their relevant parts are selected for use in further analysis; and (2) a mapping within and across high-level design and low-level design models (if available), is made. The mapped models result in a class diagram which includes the relevant elements and their relations only.

Create dependency views: In order to ensure traceability to (and between) the underlying quality models and the mapped design model, a conceptual model (a tree-formed class diagram) where classes represent elements from the underlying design and quality models, relations show the ownership, and the class attributes indicate the dependencies, interactions and the properties, is created. The conceptual model is transformed into a generic DV – a directed tree representing relationships among quality aspects and design of the system. For each quality attribute defined in the quality model, a quality attribute specific DV is created, by a new instantiation of the generic DV. Each set of nodes having a common parent is supplemented with an additional node called "Other", for completeness purpose. In addition, a total quality DV is deduced from the quality models. The DV parameters are evaluated by providing the estimates on the arcs and the leaf nodes, and propagating them according to a pre-defined model.

2.2 Verification of Prediction Models

The set of preliminary prediction models developed during the "Target modeling" phase, consists of design models, quality models and the DVs. The "Verification of prediction models" phase (Figure 3) aims to validate the prediction models (with respect to the structure and the individual parameters), before they are applied.

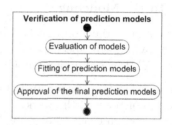

Fig. 3. Verification of models – phase

Evaluation of models: A measurement plan with the necessary statistical power is developed, describing what should be evaluated, when and how.

Fitting of prediction models: Model fitting is conducted in order to adjust the DV parameters to the evaluation results.

Approval of the final prediction models: The objective of this sub-phase is to evaluate the prediction models as a whole and validate that they are complete, correct and mutually consistent after the fitting. If the deviation is above the acceptable threshold after the fitting, the target modelling is re-initiated for structural model changes.

2.3 Application of Prediction Models

During the "Application of prediction models" phase (depicted by Figure 4), a specified change is applied and simulated on the approved prediction models.

Specify a change: The change specification should clearly state all deployment relevant facts, necessary for applying the change.

Apply the change on prediction models: This phase involves applying the specified architecture design change on the prediction models. See [12] for further details.

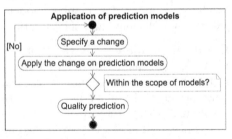

Fig. 4. Application of models – phase

Quality prediction: The propagation of the change throughout the rest of each one of the modified DVs, as well as the total quality DV, is performed based on the general DV model. See [12] for further details.

3   Application of PREDIQT in the Industrial Case

The PREDIQT method was tried out in a major industrial case study focusing on a Validation Authority (VA) [9] system.

3.1   Target Modeling

Based on the initially obtained documentation (requirement specifications, operational environment specifications etc.) and several interactions with the domain experts, UML based design models of the VA were developed.

Characterize the target and the objectives: An overview of the functional properties of the VA, the workflows it is involved in and the potential changes, was provided, in order to determine the needed level of detail and the scope of our prediction models. Among the assumed changes were increased number of requests, more simultaneous users and additional workflows that the VA will be a part of. The target stakeholder group is the system owner.

Create quality models: An extract of the quality model of the VA is shown by Figure 5. Total quality of the VA system is first decomposed into the four quality attributes: availability, scalability, security and "Other Attr." (this node covers the possibly unspecified attributes, for model completeness purpose). The X, Y, Z and V represent system quality attribute ratings, while u,v g and j represent the normalized weights expressing each attribute's contribution to the total quality of VA. The attributes and their ratings are defined with respect to the VA system, in such a manner that they are orthogonal and together represent the total quality of the VA system. Thereafter, the sub-characteristics of each quality attribute are defined with respect to the VA system, so that they are orthogonal to each other and represent a complete decomposition of the VA attribute in question. Both the attribute ratings and the sub-characteristic measures are defined so that their value lies between 0 (minimum) and 1 (maximum fulfillment).

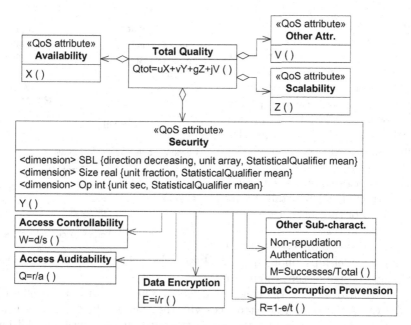

Fig. 5. An extract of the quality model

Consider the attribute "Security" in Figure 5. Its weight is v (the coefficient of Y in the "Total Quality" class). The definition of the "Security" attribute was based on [2] and its rating was based on [11]. The dimensions-notation shown in the form of attributes of the Security class in Figure 5, originates from [3]. The dimensions represent the variables constituting the rating of the Security attribute for the VA system. Given the attribute and rating definition, the appropriate (for the attribute and the VA system) sub-characteristics were retrieved from [1], where they are fully defined. The Q, W, E, R, M represent the formal measure definitions of each sub-characteristic. Security depends on the five sub-characteristics displayed: access auditability, access controllability, data encryption, data corruption prevention and "other sub-characteristics". Access auditability, for example, expresses how complete the implementation of access login is, considering the auditability requirements. Its measure is: Access auditability $= \frac{r}{a}$, where r = Nr. of information recording access log confirmed in review; a = Nr. of information requiring access log [1].

Map design models: The originally developed design models of the VA covered both high-level and low level design aspects through use cases, class diagrams, composite structure diagrams, activity diagrams and sequence diagrams. A selection of the relevant models was made, and a mapping between them was undertaken. The model mapping resulted in a class diagram containing the selected elements (lifelines, classes, interfaces etc.) as classes, ownership as relations and interactions/dependencies/properties as class attributes. Due to only ownership representing a relation, this resulted in a tree-formed class diagram.

Create dependency views: A conceptual model (a tree-formed class diagram) with classes representing the selected elements, relations denoting ownership, and selected dependencies, interactions quality properties, association relationships or their equivalents represented as the class attributes, is deduced from (1) the quality models; and (2) the above mentioned class diagram. See Figure 6 for an extract. Further details on the construction of the conceptual model are provided in [12].

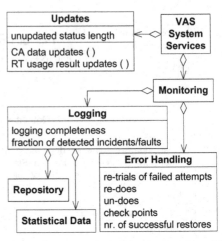

Fig. 6. An extract of the conceptual model

A generic DV – a directed tree representing relationships among quality aspects and design of the system was obtained by instantiating the conceptual model. Quality attribute specific DVs were derived in the form of directed trees with the structure from the generic DV. Each set of nodes having a common parent was supplemented with an additional node called "Other", for completeness purpose. In addition, a total quality DV was instantiated from the top two levels of the quality models.

The DVs were inserted into the tool we have built on top of MS Excel for enabling automatic simulations within and across the DVs. A QCF denotes the "degree of Quality attribute or Characteristic Fulfillment" and is associated with each node of a DV. The QCF values of attribute specific DVs were estimated by assigning a value of the quality attribute (which the DV under consideration is dedicated to) to each leaf node of the quality attribute specific DVs. The QCF value quantification involved revisiting quality models, and providing a quality attribute rating value to each node. The QCF value expresses to what degree the quality attribute (given its system specific formal definition from the quality models) represented by the DV is fulfilled within the scope of the node in question. Due to the rating definitions, the values of QCFs are constrained between 0 (minimum) and 1 (maximum). An EI denotes the "Estimated Impact" and is associated with each arc of the DVs. EI expresses the degree of impact of a child node (which the arc is directed to) on the parent node, or to what degree the parent node depends on a child node. An EI value is assigned with respect to the sub-characteristics of the quality attribute under analysis (defined in the quality models) and their respective impact on the relation in question. The EI on each arc was assigned by evaluating the impact of the child node on its parent node, with respect to the sub-characteristics (defined in the quality models) of the attribute under consideration. Once a sum of the contributions of the sub-characteristics was obtained on each arc pointing to children nodes with a common parent, the EI values were normalized so that they sum up to 1 (due to model completeness).

Figure 7 shows an extract of the Security attribute specific DV of VA (the values assigned are fictitious, for confidentiality reasons). In the case of the "Error detection" node of Figure 7, the QCF value expresses the effectiveness of error detection with respect to security. The QCFs as well as the EIs of this particular DV are estimated with reference to the definition of Security attribute and its sub-characteristics, respectively. The definitions are provided by the quality models exemplified in Figure 5. The total quality DV is assigned weights on the arcs, which, based on the attribute definitions in the quality models, express the impact of each attribute (in terms of the chosen stakeholder's gain or business criticality), on the total system quality. The weights are system general objectives. The weights are normalized and sum up to 1, since also this DV is complete. The leaf node QCFs of the total quality DV correspond to the root node QCFs of the respective quality attribute DVs. Once estimates of leaf nodes' QCFs and all EIs are provided, the QCF values of all the non-leaf nodes are automatically inferred by the tool.

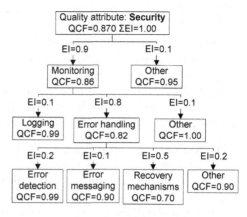

Fig. 7. A part of the Security DV

3.2 Verification of Prediction Models

Verification of the prediction models for the VA relied on measurements and expert judgments. The details are provided in Section 3.2 of [12].

3.3 Application of Prediction Models

The prediction models were applied for simulation of impacts of 14 specified, independent architecture design changes on the VA quality. Each specified architecture design change was first applied on the affected design models, followed by the conceptual model and finally the DVs. Application of a change on each quality attribute specific DV involved:

1. The DV structure was modified in order to maintain consistency with the modified conceptual model (which maps the design and the quality models).
2. For those leaf nodes that were directly traceable to the affected attributes (which represent properties, interactions and dependencies in the design models) of the conceptual model illustrated by Figure 6, the leaf node QCFs were modified by the analyst (based on the quality attribute rating definition) if appropriate for the DV in question (recall the quality model).

3. The affected arcs were identified, based on the affected attributes of the conceptual model (illustrated by Figure 6). The EI values on the affected arcs were changed by the analyst, by re-evaluating the impact of the sub-characteristics of the attribute that the DV is dedicated to and normalizing them on the arcs pointing to the nodes having a common parent. Which DVs were affected and the extent of modification of the identified EIs on them, was determined by the definitions from the quality models.
4. The modifications and their rationale were documented.

The propagation of the changes throughout and across the DVs is performed based on the general DV model, according to which the QCF value of each parent node is recursively calculated by first multiplying the QCF and EI value for each closest child and then summing these products. The DVs and the model were embedded into the above mentioned tool, which allowed simulating the change propagation in an "what-if" manner. See [12] for further details.

4   Related Work

[13] presents risk identification techniques like principal component analysis, discriminant analysis, tree-based reliability models and optimal set reduction. Common for all these techniques is that they analyze and, to some degree, predict defects, based on low-level data. [3] provides a UML notation for QoS modelling, which has been applied in our quality models. PREDIQT is also compatible with the established software quality standard [1]. The goal/question/metric paradigm [5] [4] is also compatible with PREDIQT and can be used for development of quality models and design of a measurement plan [6]. [8] and [14] introduce approaches to pattern based quantitative security assessment. Both are solely security oriented approaches for security assessment (and not prediction) with limited traceability to the system design and quality notions. [7] argues that traditional statistical approaches are inadequate and recommends holistic models for software defect prediction, using Bayesian Networks. However, a drawback that statistical and BBN-based models suffer, is poor scalability.

5   Conclusions and Future Work

This paper has presented PREDIQT – a method for model based prediction of impacts of architecture design changes on system quality. We have also reported on results from applying PREDIQT in a major industrial case study. The case study focused on prediction of security and its trade-offs with the overall quality attributes of the target system. We basically did two evaluations, the feasibility study described above and a thought experiment.

In relation to the feasibility study, a lightweight post-mortem analysis was conducted. The domain experts, who participated in the analysis, expressed that the development of DVs was relatively simple, thanks to the comprehensible DV models as well as the confined underlying quality and design models. One of the main points of the feedback was that the reasoning around DVs, particularly their parametrization, facilitated understanding the system design and reasoning about alternatives for potential improvements, as well as existing and potential weaknesses of architectural design, with respect to the quality attributes. We managed to conduct all the steps of the PREDIQT method, with limited resources and within the planned time period (six half-a-day workshops with upto one man-labour week before each). The changes specified were deduced with the objective of addressing the most relevant parts of the prediction models, being diverse and realistic. The fact that all the 14 specified changes were within the scope of the prediction models and could be simulated within the PREDIQT method, indicates feasibility of developing the prediction models with intended scope and quality. Overall, applying PREDIQT was feasible within the practical settings of this case. The models were relatively straight forward to develop, and judged to be fairly easy to use.

The predictions obtained were evaluated by means of a thought experiment on a domain expert panel with thorough knowledge of the VA system. The process and the results from simulation of the 14 specified changes in relation to the feasibility study were obtained and documented by the analysis leader, stored by an additional analysis participant, and kept unrevealed. Independently, the domain experts were asked to estimate the impact of each change on the respective quality attributes defined. The results obtained show quite a low magnitude of deviation between the PREDIQT based simulations and the values obtained from the thought experiment. We do not have hard evidence that the predictions were correct, but given the research method and the values obtained, the results of the thought experiment are promising. Further details on the PREDIQT method, the feasibility study and the thought experiment are presented in [12].

We expect PREDIQT to be applicable in several domains of distributed systems with high quality and adaptability demands. Handling of inaccuracies in the prediction models, improving traceability of the models and design of an experience factory, are among the partially initiated future developments.

Acknowledgments. This work has been conducted within the DIGIT (180052/S10) project, funded by the Research Council of Norway.

References

1. International Organisation for Standardisation: ISO/IEC 9126 - Software engineering – Product quality (2004)
2. ISO/IEC 12207 Systems and Software Engineering – Software Life Cycle Processes (2008)
3. Object Management Group: UML Profile for Modeling Quality of Service and Fault Tolerance Characteristics and Mechanisms, v. 1.1 (2008)
4. Basili, V., Caldiera, G., Rombach, H.: The Goal Question Metric Approach. In: Encyclopedia of Software Engineering. Wiley, Chichester (1994)
5. Basili, V.R.: Software Modeling and Measurement: the Goal/Question/Metric Paradigm. Technical Report TR-92-96, University of Maryland (1992)

6. Ebert, C., Dumke, R., Bundschuh, M., Schmietendorf, A., Dumke, R.: Best Practices in Software Measurement. Springer, Heidelberg (2004)
7. Fenton, N., Neil, M.: A Critique of Software Defect Prediction Models. IEEE Transactions on Software Engineering 25, 675–689 (1999)
8. Heyman, T., Scandariato, R., Huygens, C., Joosen, W.: Using Security Patterns to Combine Security Metrics. In: ARES 2008: Proceedings of the 2008 Third International Conference on Availability, Reliability and Security, Washington, DC, USA, pp. 1156–1163. IEEE Computer Society, Los Alamitos (2008)
9. Ølnes, J.: PKI Interoperability by an Independent, Trusted Validation Authority. In: 5th Annual PKI R&D Workshop, NIST Gaithersburg MD, USA (2006)
10. Ølnes, J., Andresen, A., Buene, L., Cerrato, O., Grindheim, H.: Making Digital Signatures Work across National Borders. In: ISSE 2007 Conference: Securing Electronic Business Processes, pp. 287–296. Warszawa, Vieweg Verlag (2007)
11. Omerovic, A.: Design Guidelines for a Monitoring Environment Concerning Distributed Real-Time Systems. Tapir Academic Press, Trondheim (2004)
12. Omerovic, A., Andresen, A., Grindheim, H., Myrseth, P., Refsdal, A., Stølen, K., Ølnes, J.: A Feasibility Study in Model Based Prediction of Impact of Changes on System Quality. Technical report, SINTEF A13339 (2009)
13. Tian, J.: Software Quality Engineering: Testing, Quality Assurance, and Quantifiable Improvement, 1st edn. Wiley-IEEE Computer Society Press, Chichester (2005)
14. Yautsiukhin, A., Scandariato, R., Heyman, T., Massacci, F., Joosen, W.: Towards a Quantitative Assessment of Security in Software Architectures. In: 13th Nordic Workshop on Secure IT Systems, Copenhagen, Denmark (2008)

Author Index